WITHDRAWN
University of
Illinois Library
at Urbana-Champaign

Polymer Characterization
ESR and NMR

Polymer Characterization by ESR and NMR

Arthur E. Woodward, EDITOR
City College of New York

Frank A. Bovey, EDITOR
Bell Laboratories

Based on a symposium sponsored by the Division of Polymer Chemistry at the 178th Meeting of the American Chemical Society Washington, D.C., September 10–14, 1979.

ACS SYMPOSIUM SERIES 142

AMERICAN CHEMICAL SOCIETY
WASHINGTON, D. C.　1980

Library of Congress CIP Data
Polymer characterization by ESR and NMR.
(ACS symposium series; 142 ISSN 0097–6156)

Based on a symposium sponsored by the Division of Polymer Chemistry at the 178th meeting of the American Chemical Society, Washington, D.C., September 10–14, 1979.
Includes bibliographies and index.

1. Polymers and polymerization—Congresses. 2. Electron paramagnetic resonance—Congresses. 3. Nuclear magnetic resonance—Congresses.
I. Woodward, Arthur E., 1925– . II. Bovey, Frank Alden, 1918– . III. American Chemical Society. Division of Polymer Chemistry. IV. Series: American Chemical Society. ACS symposium series; 142.

QD471.P68   547.8'4046   80–21840
ISBN 0-8412-0594-9 ACSMC8 142 1–309 1980

Copyright © 1980

American Chemical Society

All Rights Reserved. The appearance of the code at the bottom of the first page of each article in this volume indicates the copyright owner's consent that reprographic copies of the article may be made for personal or internal use or for the personal or internal use of specific clients. This consent is given on the condition, however, that the copier pay the stated per copy fee through the Copyright Clearance Center, Inc. for copying beyond that permitted by Sections 107 or 108 of the U.S. Copyright Law. This consent does not extend to copying or transmission by any means—graphic or electronic—for any other purpose, such as for general distribution, for advertising or promotional purposes, for creating new collective works, for resale, or for information storage and retrieval systems.

The citation of trade names and/or names of manufacturers in this publication is not to be construed as an endorsement or as approval by ACS of the commercial products or services referenced herein; nor should the mere reference herein to any drawing, specification, chemical process, or other data be regarded as a license or as a conveyance of any right or permission, to the holder, reader, or any other person or corporation, to manufacture, reproduce, use, or sell any patented invention or copyrighted work that may in any way be related thereto.

PRINTED IN THE UNITED STATES OF AMERICA

ACS Symposium Series

M. Joan Comstock, *Series Editor*

Advisory Board

David L. Allara

Kenneth B. Bischoff

Donald G. Crosby

Donald D. Dollberg

Robert E. Feeney

Jack Halpern

Brian M. Harney

Robert A. Hofstader

W. Jeffrey Howe

James D. Idol, Jr.

James P. Lodge

Leon Petrakis

F. Sherwood Rowland

Alan C. Sartorelli

Raymond B. Seymour

Gunter Zweig

FOREWORD

The ACS SYMPOSIUM SERIES was founded in 1974 to provide a medium for publishing symposia quickly in book form. The format of the Series parallels that of the continuing ADVANCES IN CHEMISTRY SERIES except that in order to save time the papers are not typeset but are reproduced as they are submitted by the authors in camera-ready form. Papers are reviewed under the supervision of the Editors with the assistance of the Series Advisory Board and are selected to maintain the integrity of the symposia; however, verbatim reproductions of previously published papers are not accepted. Both reviews and reports of research are acceptable since symposia may embrace both types of presentation.

CONTENTS

Preface ix

1. Conformation and Mobility of Polymers Adsorbed on Oxide Surfaces by ESR Spectroscopy 1
   Tai Ming Liang, Peter Dickenson, and Wilmer G. Miller

2. An ESR Investigation of Environmental Effects on Nylon Fibers .. 19
   K. L. DeVries and M. Igarashi

3. An ESR Study of Initially Formed Intermediates in the Photodegradation of Poly(vinyl chloride) 35
   N.-L. Yang, J. Liutkus, and H. Haubenstock

4. Manganese(II) and Gadolinium(III) EPR Studies of Membrane-bound ATPases 49
   Charles M. Grisham

5. Molecular Structure of Vinyl Chloride–Vinylidene Chloride Copolymers by Carbon 13 NMR 81
   Charles J. Carman

6. Characterization of Long-Chain Branching in Polyethylenes Using High-Field Carbon-13 NMR 93
   J. C. Randall

7. Molecular Dynamics of Polymer Chains and Alkyl Groups in Solution 119
   George C. Levy, Peter L. Rinaldi, James J. Dechter, David E. Axelson, and Leo Mandelkern

8. Use of Pulsed NMR to Study Composite Polymeric Systems 147
   D. C. Douglass

9. Multiple-Pulse NMR of Solid Polymers: Dynamics of Polytetrafluoroethylene 169
   A. D. English and A. J. Vega

10. Variable-Temperature Magic-Angle Spinning Carbon-13 NMR Studies of Solid Polymers 193
    W. W. Fleming, C. A. Fyfe, R. D. Kendrick, J. R. Lyerla, H. Vanni, and C. S. Yannoni

11. Drug–DNA Interactions in Solution: Acridine Mutagen, Anthracycline Antitumor, and Peptide Antibiotic Complexes 219
    Dinshaw J. Patel

Index 295

PREFACE

A basic requirement of the ESR technique is the presence of molecules or atoms containing unpaired electrons. Such species can be generated in polymeric systems by homolytic chemical scission reactions or by polymerization processes involving unsaturated monomers. These reactions can be initiated thermally, photochemically, or with a free-radical initiator, and, in the case of scission, by mechanical stress applied to the system. Therefore, ESR can be used to study free-radical-initiated polymerization processes and the degradation of polymers induced by heat, light, high-energy radiation, or the application of stress.

Molecules containing unpaired electrons also can be introduced into a polymeric system in a number of other ways. A polymer molecule can be *spin labelled* by chemically reacting it with a small molecule containing an unpaired electron in a relatively stable state. Transition metal ions can be added by substitution reactions on the polymer molecules or can be already present as part of the polymer molecules. One also can add a *spin probe,* a small molecule with an unpaired electron in a relatively stable state; spin probes are held in the polymeric matrix by physical forces. The ESR spectra of spin labels, spin probes, and transition metal ions have been used to study a variety of phenomena associated with polymeric systems including: the molecular motion and relaxation processes of polymers in solution and in the rubbery and partially crystalline states; the adsorption of polymers onto solid surfaces; binding and binding-site locations in proteins and other biological macromolecules; and conformational changes in polymers and in lipid–membrane systems.

The NMR technique is a highly versatile one, and has been applied in many forms to the study of macromolecules. Examples of all of these are provided in the contributions to this volume.

Perhaps the widest application is that of conventional high-resolution spectroscopy in solution for the purpose of learning in detail about polymer chain structure. In this field, proton NMR, formerly dominant, has given way to carbon-13 NMR with the development of pulse Fourier transform spectrometers with spectrum accumulation. Carbon spectroscopy is capable of giving very detailed and often quite sophisticated information. For example, a very complete accounting can be provided of comonomer sequences in vinyl copolymers; and branches can be identified and counted, even at very low levels, in polyethylenes.

The polymer application having the longest history (nearly as long as that of NMR itself) is its observation in the solid state. Here, because of magnetic dipole interactions (which are averaged to zero in solution), resonances are very broad (of the order of kilohertz) and detailed structural information is ordinarily entirely lost. Interest centers instead on nuclear relaxation: spin-lattice relaxation, measured by T_1; spin–spin relaxation, T_2; and spin-lattice relaxation in the rotating frame, $T_{1\rho}$. From the interpretation of these quantities, particularly their temperature dependence, important conclusions concerning the motions of the chains can be drawn, and from this knowledge in turn one can obtain insights into the nature of composite macromolecular systems such as polymer–plasticizer and polymer–polymer blends and partially crystalline polymers. By using appropriate multiple sequences of pulses, chemical shift information may be recovered and macromolecular motions measured in separate domains of composite systems.

For dilute spins such as carbon-13, chemical shift information may be obtained by cross-polarization combined with high-power decoupling to remove dipole–dipole broadening. The carbon resonances thus obtained are still broadened owing to chemical shift anisotropy. By rapid (kilohertz) spinning at the "magic angle," the latter may be averaged out and the results approximate to solution spectra, although with considerable line broadening. Thus cross-linked and insoluble polymers may be examined, and it is also possible to obtain information concerning the motions of individual carbons.

Finally, high-resolution NMR has proved to be a technique of extraordinary power in the examination of the detailed structure of biological macromolecules, principally proteins and nucleic acids. It usefully complements their study in the crystalline state by X-ray diffraction, and while it cannot be said precisely to rival X-ray it also is capable of supplying many hundreds of structural parameters. In addition, NMR can provide many insights that X-ray cannot, including kinetic information.

FRANK A. BOVEY
Bell Laboratories
Murray Hill, NJ 07974
July 10, 1980

ARTHUR E. WOODWARD
City College of New York
New York, NY 10031

Conformation and Mobility of Polymers Adsorbed on Oxide Surfaces by ESR Spectroscopy

TAI MING LIANG, PETER N. DICKENSON, and WILMER G. MILLER

Department of Chemistry, University of Minnesota, Minneapolis, MN 55455

The nature of the interaction of polymers with solid surfaces is of much practical importance, e.g., in coatings and in reinforced polymers, as well as of theoretical interest (1,2,3). Most experimental and theoretical studies at the molecular level have been concerned with the adsorbed polymer at a solid-liquid interface. The interaction of the dry polymer with the solid surface, of significance in applied problems, is little understood at the segmental or even molecular level. Instead, interaction is generally inferred from effects on bulk properties.

Theoretical treatments (e.g., 4-15) of adsorbed polymers at the solid-liquid interface predict some or all of the following properties: average fraction of segments adsorbed; average length of loops, adsorbed trains and tails; mean extension of segments above the surface; effect of polymer-solvent and polymer-surface interaction; effect of molecular weight. Experimentally the average thickness of the adsorbed layer upon drying is estimated from adsorption isotherm-surface area measurements, and at the solid-liquid interface by viscosity, or by ellipsometry. Under favorable circumstances the fraction of segments adsorbed is determined by infrared spectroscopy. Little is known about the mobility of segments in loops and tails, whether in the presence or absence of a contacting solvent. In this communication we discuss the use of a stable nitroxide free radical and ESR spectroscopy to monitor segmental dynamics, and summarize our efforts (16,17,18,19,20). Previous studies, less comprehensive in scope, have been made by others (21,22,23,24).

Basis of Method

The line shape of the ESR spectrum of a nitroxide free radical in x-band operation varies with the rotational motion of the nitroxide over the range 10^{-11} to 10^{-6} sec in rotational correlation time. By saturation transfer measurements even slower motions may be studied (25). The rotational correlation time for a nitroxide in a dilute solution of labelled random coil polymer in a low viscosity solvent falls typically in the 0.05 to 0.5 nsec range, giving a three line spectrum as shown in Figure 1A. The relaxation of the nitroxide is facilitated by backbone motions extending only a few atoms from the point of attachment, i.e., to local mode motions, as deduced from molecular weight and other studies by several investigators (26,27). Thus in an adsorbed molecule an isolated loop or tail extending into the solvent should give a motionally narrowed three line spectrum if the loop or tail is greater than a minimum size (probably four-to-six bonds about which rotation can occur). The spectral line shape is concentration dependent, but remains motionally narrowed until relatively high polymer concentrations are reached (Figure 1-B,C and reference 27). Loop and tail segment density near the surface may be large, but falls off rapidly with distance above the surface. We therefore expected tails and all but the smallest loops (and surface extended loops) would exhibit three line spectra similar to that observed with the free, isolated polymer molecule.

By contrast a nitroxide attached to monomeric units rigidly held to a solid surface should show a spectral line shape similar to the bulk polymer in the glassy state (Figure 1E). Thus a polymer molecule adsorbed at a solid-liquid interface should exhibit a composite spectrum, from which one can deduce the fraction of the monomeric units in loops and tails, and their mean motion. The sensitivity of the method to small amounts of units with motional freedom can be seen from Figure 2.

A final point to consider is the effect of nonbound polymer. When a surface is added to a polymer solution and equilibrium is reached, both adsorbed and free molecules are present. Although adsorbed and free polymer are in dynamic equilibria, desorption is generally sufficiently slow that the nonbound polymer can be separated and monitored independently of the bound polymer.

In the absence of solvent, the influence of the surface on the segmental mobility can be deduced by measuring the temperature dependence of the spectral line shape in the presence and absence of the surface. The temperature dependence of bulk poly(vinyl acetate) is shown in Figure 3, and of bulk polystyrene in Figure 10. Although the polymer-surface interaction may have a profound effect on segmental mobility, it seems unlikely that any composite spectra can be decomposed easily into bound and unbound contributions.

Experimental

Poly(vinyl acetate), PVAc, of molecular weights (M_w) 6.1×10^4, 1.9×10^5, and 6.0×10^5, was randomly labeled by ester exchange with 2,2,5,5-tetramethyl-3-pyrrolin-1-oxyl-3-carboxylic acid to give a spin labeled polymer containing typically 1-10 nitroxides per polymer molecule. Polystyrene was prepared by emulsion

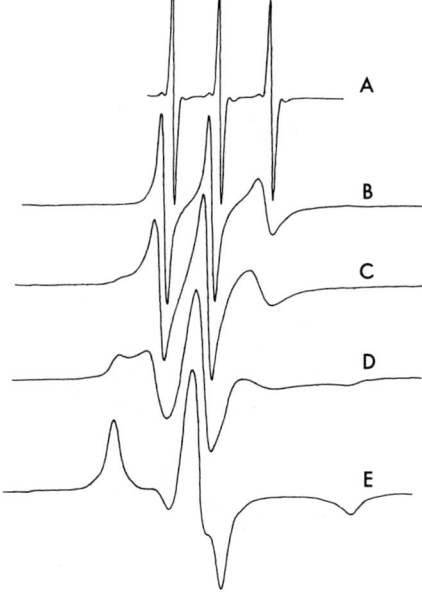

Figure 1. ESR spectrum at 25°C as a function of concentration of PVAc-(61,000), randomly labeled. Polymer concentration (weight percent) in $CHCl_3$ is (A) 1%; (B) 44%; (C) 61%; (D) 72%; and (E) 100% (bulk).

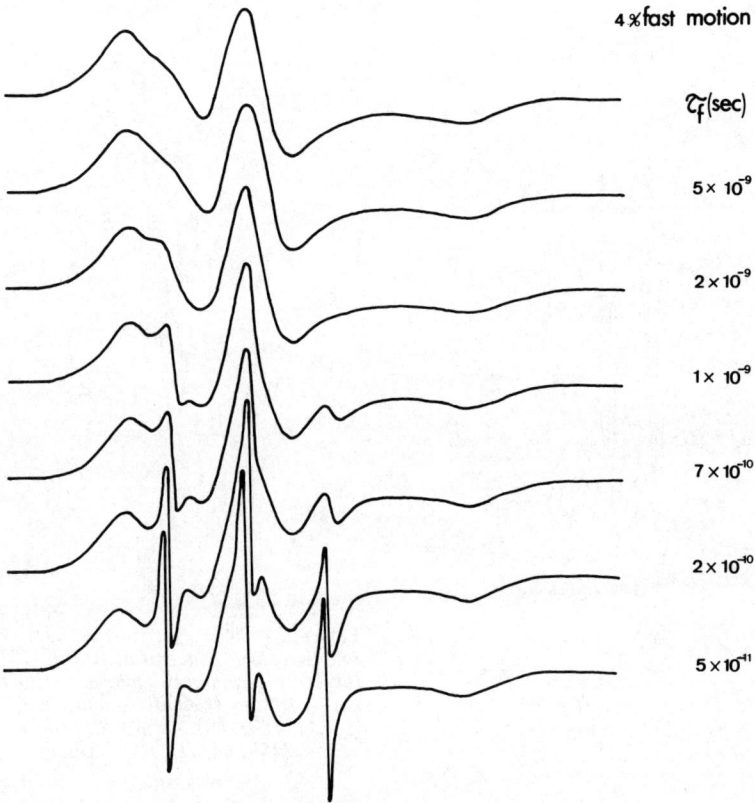

Figure 2. Simulated composite spectra consisting of 4% of the nitroxides undergoing motional narrowing (isotropic, homogeneous broadening of 1 G) with rotational correlation times as indicated

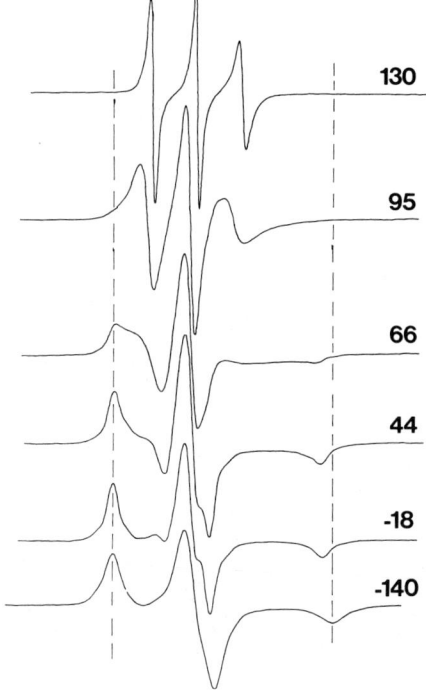

Figure 3. Temperature dependence of bulk PVAc. The glass transition is ~ 30°C. (Temperatures are in Celsius.)

polymerization of styrene with 2 wt. % chloromethyl styrene, followed by partial reaction of the chloride with 2,2,6,6-tetramethyl-4-piperidinol-1-oxyl (28).

Alumina, Al_2O_3, obtained from Matheson, Coleman and Bell, was heated at 200°C for two hrs. before use. Titanium dioxide was obtained from Polysciences as 0.45 μ spheres. Before use it was treated to remove paramagnetic impurities (19). Glass (soda lime) spheres of 3-8 μ nominal diameter was obtained from Polysciences and treated to remove paramagnetic impurities. SiO_2 in the form of aggregated 0.014 μ diameter spheres (Cab-O-Sil M5) was used as received.

For adsorption studies the labeled polymer was dissolved in reagent grade solvent, the oxide surface added, and the sample was stirred magnetically for at least 24 hrs. (sometimes up to 3 days). The oxide with adsorbed polymer was separated from the unadsorbed polymer by gravity settling, or where necessary by low field centrifugation. The oxide together with adsorbed polymer was washed with fresh solvent until no ESR activity was detectable in the supernatant. The ESR spectrum was then taken of the polymer on the surface either in the presence of solvent, or after solvent removal as a function of temperature.

The ESR spectra were measured on a Varian E-3 spectrometer at about 9.15 GHz. The spectra were typically recorded in the vicinity of 3.2 kG with attenuation power low enough to avoid saturation.

Conformation and Mobility at the Solid-Liquid Interface

Effect of Surface. The spectrum of PVAc(61,000) adsorbed at a liquid ($CHC\ell_3$)-solid interface at full surface coverage is shown in Figure 4. Chloroform is a thermodynamically good solvent for PVAc, estimated to be nearly an athermal solvent (29). The differences are quite striking, ranging from $A\ell_2O_3$, where all units are rigidly held, to Cab-O-Sil SiO_2, where the majority of the spin labels are in flexible loops and tails. The immobilization of the spin labels is through the side chain ester, and not through the nitroxide moiety, as has been shown through use of spin probes of varying functionality (17).

Changing the contacting solvent strongly affects the fraction of immobilized units, but the effect of changing the surface in contact with a given solvent shows a trend analogous to the chloroform results. The ranking of the surfaces in order of decreasing interaction with PVAc is $A\ell_2O_3$, TiO_2, glass(3-8 μ soda lime), and SiO_2(0.014 μ aggregated spheres.)

Effect of Solvent. With PVAc adsorbed at the $A\ell_2O_3$-liquid interface, no motionally narrowed component is observed with any solvent, ranging from nearly athermal ($CHC\ell_3$) to nearly a theta solvent ($CC\ell_4$) (17,30). With PVAc(61,000) at the TiO_2-liquid interface, as shown in Figure 5, a small percentage (<5) of very

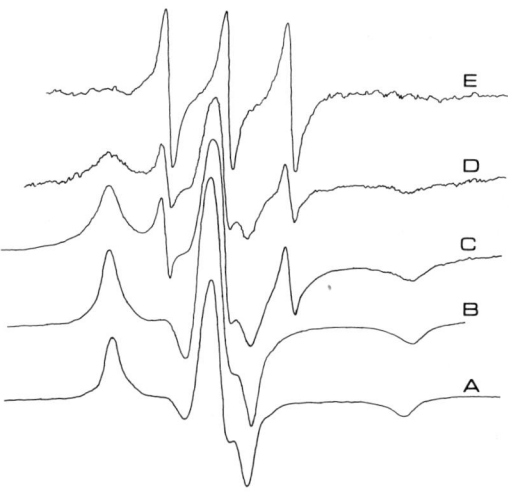

Figure 4. The spectra of PVAc (61,000) adsorbed at the solid–liquid (CHCl₃) interface at saturation surface coverage: (A) no solvent or surface; (B) Al_2O_3; (C) TiO_2; (D) glass (3–8 μ); (E) SiO_2 (Cab-O-Sil M5).

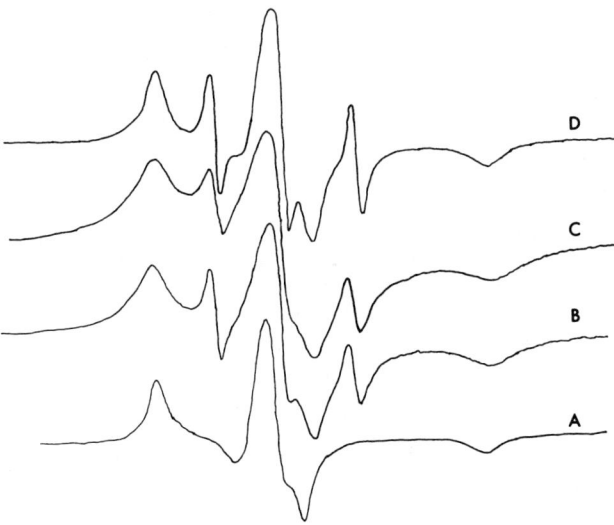

Figure 5. The ESR spectra of PVAc (61,000) at the solid (TiO_2)–liquid interface at saturation coverage: (A) no solvent or surface; (B) CCl_4; (C) toluene; (D) $CHCl_3$.

mobile nitroxides is observed. There is not a strong trend with solvent, though over a series of experiments (19) thermodynamically good solvents lead to a slightly greater percentage of fast component than poor solvents. The most striking change is with SiO_2 (Cab-O-Sil), where going from poor to good solvents leads to a dramatic increase in the amount of the fast component (19). Alternatively, raising the temperature is thermodynamically equivalent to changing to a better solvent. The results are consistent, in that raising the temperature leads to an increase in mobile units for any solvent-surface pair examined (19).

The sharpness of the fast component in the composite spectra in Figures 4 and 5 is a good indication that there is not a wide distribution of correlation times, and by reference to Figure 1, that the mobile segment density of the adsorbed polymer extending into the solvent is dilute. The nature of the mobile units-loops, tails, or some of each-cannot be deduced from these data.

Effect of Molecular Weight. The nature of the mobile units can best be probed by a study of the molecular weight dependence. Shown in Figure 6 is the molecular weight dependence, at saturation surface coverage, of PVAc on TiO_2 in the presence of chloroform. Although the molecular weight is varied by a factor of ten, the fraction of fast component is unchanged, within experimental error. Similar behavior is observed in other surface-solvent systems (19).

All theoretical treatments of the molecular weight dependence indicate the number of monomeric units in loops increases monotonically with molecular weight, whereas the number of units in tails approaches an asymptotic limit, i.e., the fraction of units in tails must be a decreasing function of molecular weight. The molecular weight invariance we observe indicates the mobile units are predominantly in loops, and not tails.

Effect of Surface Coverage. The effect of surface coverage is illustrated in Figure 7 with PVAc at the SiO_2-$CC\ell_4$ interface. The effect is similar with other solvents, and with the glass and TiO_2 surfaces. As the surface coverage increases the fraction of units in loops increases, in some cases quite dramatically as, for example, in Figure 7. This result was inadvertently stated incorrectly in reference 18. At low surface coverage the polymer lies close to the surface in a flattened conformation. Monomeric units in any loops have little motional freedom, irrespective of size. As the surface coverage increases a less flattened configuration is assumed. These observations are qualitatively similar to those for poly(vinyl pyrrolidone) adsorbed on aerosil SiO_2 (21-24).

Effect of Polymer. The effect of changing the polymer while keeping the surface and solvent fixed is illustrated by comparing Figures 4 and 8. There are two factors which must be considered:

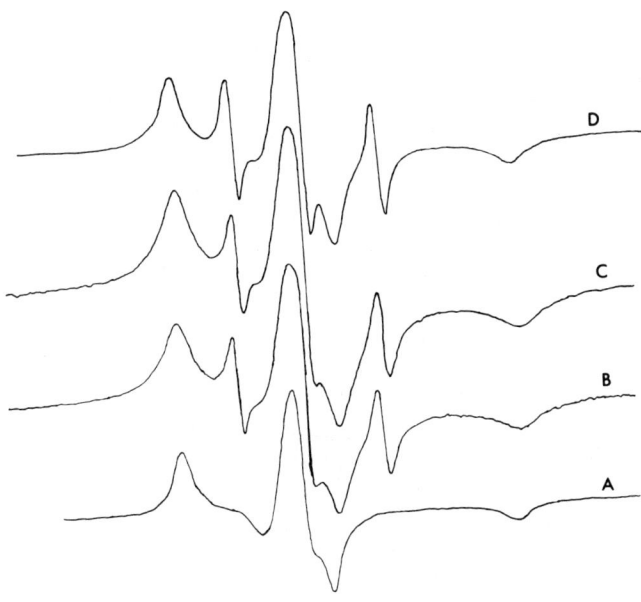

Figure 6. Effect of molecular weight of PVAc on TiO_2 in the presence of $CHCl_3$ at saturation coverage: (A) no solvent or surface; (B) 61,000; (C) 194,000; (D) 600,000.

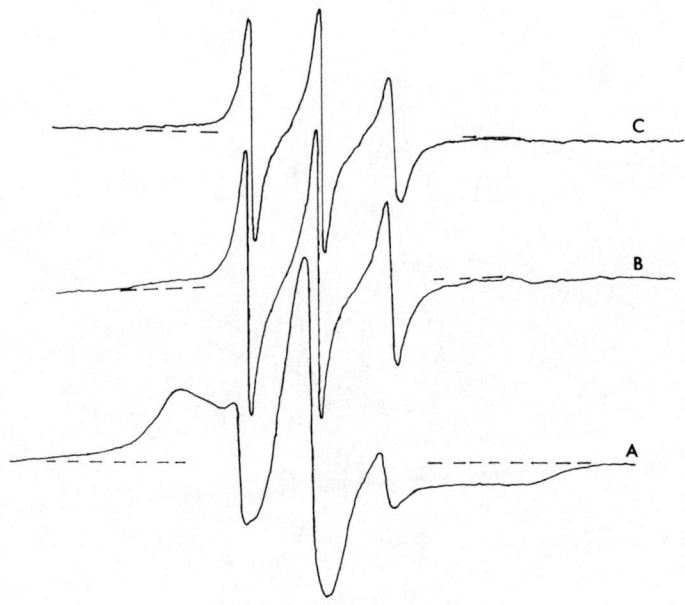

Figure 7. The dependence of the ESR spectra on surface coverage of PVAc (61,000) at the SiO_2(Cab-O-Sil)–CCl_4 interface at 25°C. Percent of saturation uptake: (A) 50%; (B) 90%; (C) at saturation.

polymer-solvent and polymer-surface interactions. Thermodynamically $CHCl_3$ should be a poorer solvent for polystyrene than for PVAc (29). Therefore, one would expect less fast component with polystyrene. However, we observe a very low amount of slow component with polystyrene at either the SiO_2-$CHCl_3$ or the glass-$CHCl_3$ interface. This must reflect a difference in the interaction of the styrene and vinyl acetate units with the surface.

Effect of the Surface on Segmental Mobility in the Dry State

The Polymer-Surface Interaction. When using a nitroxide labeled polymer the polymer-surface interaction in the absence of solvent can be assessed from temperature studies in the presence and absence of the surface, as shown in Figures 9 and 10. At temperatures below or near the glass transition the effect of the surface cannot be deduced as all motions are too slow. However, as the temperature is raised above T_g the effect of the surface is clearly discernible, with surface inhibiting segmental motion. It is evident from Figures 9 and 10 that the absolute temperature is not as important as $T-T_g$, where T_g is the T_g in the absence of the surface. The effect of the $T_{g\,surface}$ is still discernible at temperature $\sim$100° above T_g. As the temperature is further increased the segmental mobility becomes independent of the presence of the surface (20), indicating that kT has become greater than the segment-surface interaction energy. However, it was found that the spectrum observed at any temperature was independent of the thermal history of the sample. Inasmuch as the samples were prepared by evaporating the solvent at room temperature, there was no *a priori* reason to expect such behavior. This suggests that as the solvent is evaporated the polymer conformation on the surface approaches the equilibrium conformation, as it does when the dry surface adsorbed polymer is cooled from high temperatures, where it is effectively desorbed.

Similar studies (19) with PVAc adsorbed on the glass spheres indicates a smaller fraction of mobile units at any temperature compared to the SiO_2 studies. Since the composition of the Cab-O-Sil SiO_2 and the soda lime glass differ, it is tempting to ascribe this to differences in segment-surface interaction. There is, however, another possibility due to the difference in surface morphology. The PVAc has a radius of gyration ranging from $\sim$70 to over 200 Å, depending on the molecular weight and presence or absence of solvent. The 3-8 μ glass spheres, with a radius over 100 times the radius of gyration of the polymers, present effectively a flat surface to the polymer. By contrast the Cal-O-Sil SiO_2, composed of clusters of 70 Å radius spheres, present a highly curved surface to the polymer chain. It is clear that many chains are associated with more than one cluster as three dimensional networks are formed. The observed differences may be then a morphological one rather than a difference in segment-surface interaction.

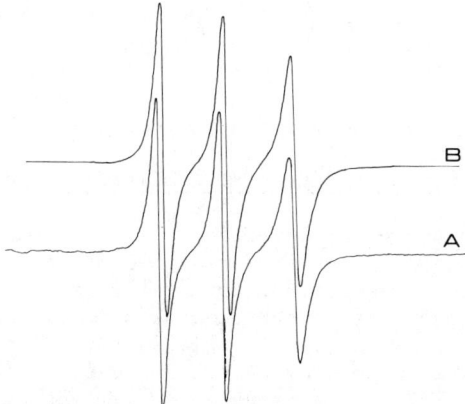

Figure 8. The effect of changing the polymer while keeping the surface and solvent fixed. PS at (A) the $CHCl_3$–SiO_2 interface; (B) the $CHCl_3$–glass interface.

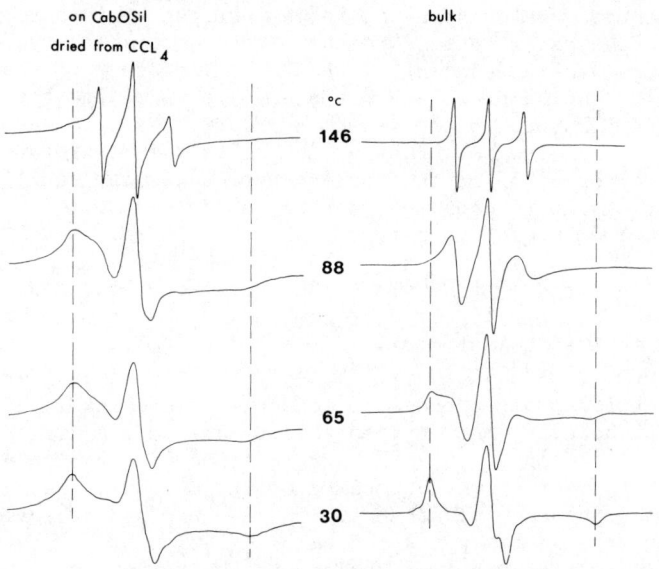

Figure 9. Temperature dependence of PVAc (6×10^5) in the bulk state and when adsorbed onto SiO_2 (obtained by drying a sample adsorbed from CCl_4 at 50% surface coverage)

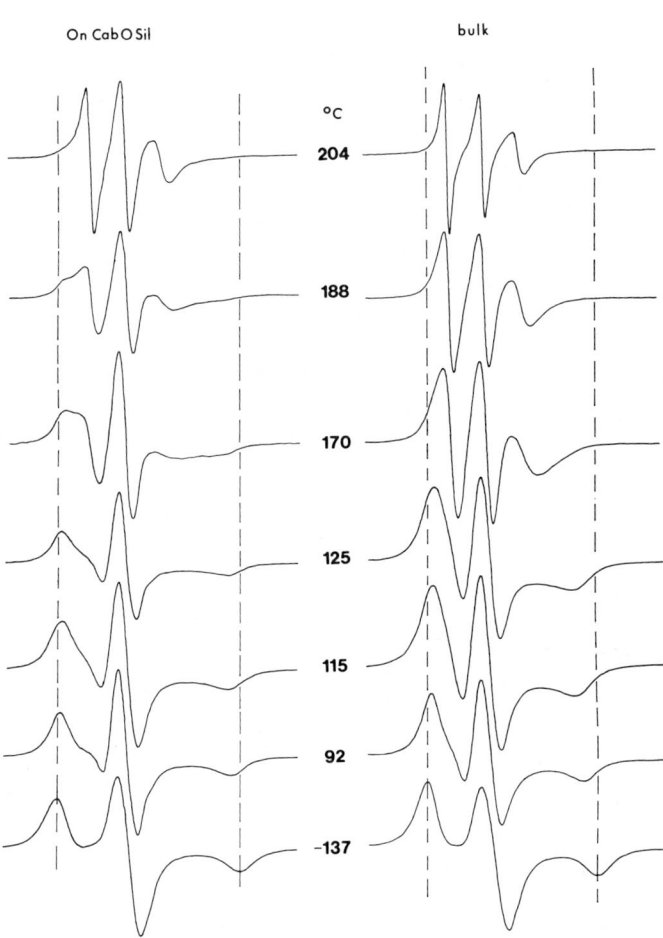

Figure 10. Temperature dependence of PS in the bulk state, and when adsorbed onto SiO_2 (obtained by drying a sample adsorbed from CCl_4). The glass transition for bulk PS is ~ 100°C.

Lateral Dependence. In the presence of solvent, loops were found to have segmental mobility similar to that of non-adsorbed polymer. It was thus of interest to deduce how segmental mobility of loops in the solid state were affected by the surface. Two types of experiments were designed to provide an answer, examples of which are shown in Figures 11 and 12. At higher surface coverage the mean thickness in the dry state is larger, hence more polymer segments are involved in loops. It is clear from Figures 11, 9 and 3 that at saturation surface coverage (dried from CCl_4) most of the spin labels have mobilities similar to that in the bulk polymer in the absence of a surface. However, in Figure 12 we see that at saturation surface coverage, when dried from $CHCl_3$, there is a considerable effect of the surface. This is easily explained. The polymer uptake is solvent dependent. In fact, at 50% saturation in CCl_4 the polymer uptake is nearly identical to saturation coverage in CCl_4 (19). Thus when the solvent is removed the thickness of the polymer coat is the same upon drying a sample 50% saturated from CCl_4 as at saturation coverage from $CHCl_3$, and should behave similarly if equilibrium conformation is achieved. Comparison of the appropriate spectra (Figure 12 and reference 20) shows this to be true. If one takes the surface area of the Cab-O-Sil ($\sim$200 m^2/gm) and assumes that the entire surface is equally accessible to the polymer, the mean polymer thickness under the two conditions is $\sim$6 Å and $\sim$12 Å. From these results one can state that the surface does not affect the mobility of a segment when it is more than a few angstroms from the surface.

Segmental Dynamics. Two types of dynamics can be considered-segmental motion, and diffusion of the polymer from the surface. Previously we have seen that at the solid-liquid interface segmental motion in loops appears to differ little from that in dilute, dissolved polymer, unless the loops are held close to the polymer surface. In the dry state temperature studies indicated that segmental motion of loops differed little from segmental motion in bulk polymer.

Inasmuch as segmental motion above T_g becomes rapid, the diffusion of the polymer off the surface can be monitored, an example of which is shown in Figure 13. A sample of labeled PVAc was surface adsorbed ($\sim$50% saturation coverage) and dried. It was then blended at -80 °C with unlabeled PVAc, and compression molded at 9000 psi at 30 °C. The sample was rapidly raised to 146 °C and monitored. By the time the first spectrum was obtained ($\sim$8 minutes) all segments had been desorbed and replaced by overcoated, unlabeled polymer, i.e., the labeled polymer had diffused from the surface into the bulk polymer. Although this seemed at first to be surprising, when one considers that the

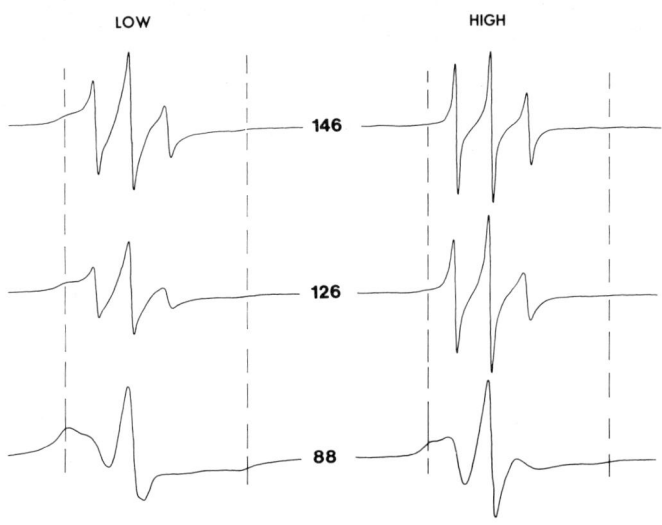

Figure 11. Temperature dependence of PVAc (6×10^5) on SiO_2 when adsorbed from CCl_4 to 50% of saturation coverage (low) before drying, or at saturation coverage (high) before drying

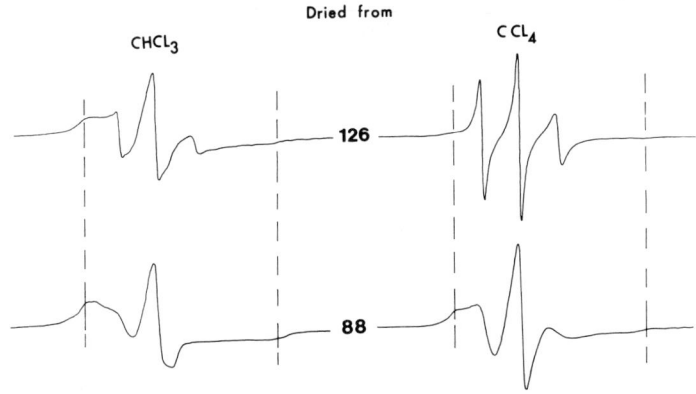

Figure 12. Comparison of the temperature dependence of PVAc (6×10^5) on SiO_2 when adsorbed at saturation coverage from CCl_4 and from $CHCl_3$ before drying

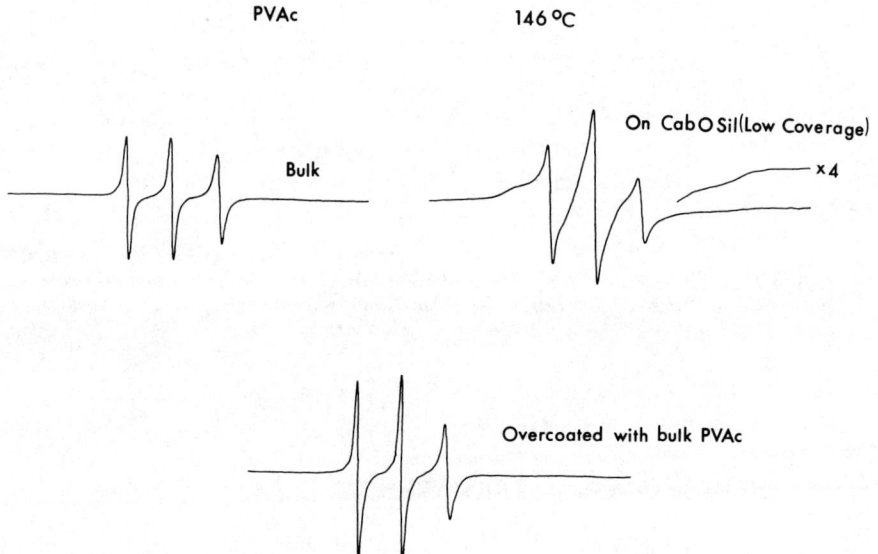

Figure 13. Diffusion of PVAc from SiO$_2$ surface into bulk polymer (see text for details)

thickness of the labeled polymer was less than 10 Å, the time scale seems more reasonable.

Discussion

The adsorption of PVAc and polystyrene has been studied previously (1,2). With PVAc at glass-**toluene** and glass-benzene interfaces, thicknesses ranging from hundreds to thousands of angstroms have been reported (31,32,33,34). In these polymer-solvent-surface systems we would conclude that the mean thickness was at most tens of angstroms, while in the $PVAc-CHCl_3-SiO_2$, $PVAc-CCl_4-SiO_2$, polystyrene-$CHCl_3-SiO_2$ and $PVAc-CHCl_3$-glass systems at saturation, thicknesses approaching the polymer radius of gyration are compatible with the data.

The nitroxide labeling technique seems an excellent approach for exploring a number of factors concerned with polymer conformation and dynamics at the solid-liquid interface, and with the adsorbed polymer in the absence of solvent. It is difficult to compare many of the results with theoretical predictions, as a numerical value for the differential adsorption energy parameter is difficult to determine. Also many theoretical predictions pertain to the infinite chain and the isolated molecule. The recent theory of Scheutjens and Fleer (15) takes into account many factors of experimental importance, and is presented in a form which can be qualitatively compared with experiment. For high chain length (r=1000) the fraction of nonbound segments is predicted to increase with surface coverage, and to be greater for athermal solvents than for theta solvents. We observe these trends. However, we do not find evidence for long, dangling tails, as predicted under some circumstances.

Our measurements on the dynamics of dry, adsorbed polymer has no known theoretical counterpart.

Acknowledgements

This work was supported by the Petroleum Research Fund (8749-AC5,6), administered by the American Chemical Society, NIH (GM 16922), and NSF (undergraduate summer fellowship to PND).

Literature Cited

1. Patrick, R. L., Ed. "Treatise on Adhesion and Adhesives"; Marcel Dekker: New York, 1967.
2. Lipatov, Yu. S.; Sergeeva, L. M. "Adsorption of Polymers"; Halsted Press: New York, 1974.
3. Mittal, K. L., Ed. "Adsorption at Interfaces"; Amer. Chem. Soc.: Washington, D.C.; 1975.
4. Simha, R.; Frisch, H. L.; Eirich, F. R. *J. Phys. Chem.*, 1953, 57, 584.

5. Silberberg, A. J. Phys. Chem., 1962, 66, 1872.
6. DiMarzio, E. A. J. Chem. Phys., 1965, 42, 2101.
7. Rubin, R. J. J. Chem. Phys., 1965, 43, 2392.
8. Roe, R. J. J. Chem. Phys., 1965, 43, 1591.
9. Motomura, K. Matuura, R.; J. Chem. Phys., 1969, 50, 1281.
10. Silberberg, A. J. Chem. Phys., 1968, 48, 2835.
11. Hoeve, C.A.J. J. Polym. Sci. C., 1970, 30, 361.
12. Hoeve, C.A.J. J. Polym. Sci. C., 1971, 34, 1.
13. Roe, R. J. J. Chem. Phys., 1974, 60, 4192.
14. Helfand, E. Macromolecules, 1976, 9, 307.
15. Scheutjens, J.M.H.M.; Fleer, G.J. J. Phys. Chem., 1979, 83, 1619.
16. Miller, W. G.; Veksli, Z. Rubber Chem. and Tech., 1975, 48, 1978.
17. Miller, W. G.; Rudolf, W. T.; Veksli, Z.; Coon, D. L.; Wu, C. C.; Liang, T. M. in MMI Press Symposium Series, Vol. 1, "Molecular Motion in Polymers by ESR"; Harwood Academic Publ. GmbH: Chur, Switzerland, 1979.
18. Miller, W. G.; Liang, T. M. Polymer Preprints, 1979, 20(2), 189.
19. Liang, T. M.; Miller, W. G. J. Colloid Interface Sci., xxxx.
20. Liang, T. M.; Tan, S. W.; Dickson, P. N.; Miller, W. G. J. Colloid Interface Sci., xxxx.
21. Fox, K. K.; Robb, I. D.; Smith, R. J. Chem. Soc. Faraday Trans. I., 1974, 70, 1186.
22. Robb, I. D.; Smith, R. Eur. Polym. J., 1974, 10, 1005.
23. Clark, A. T.; Robb, I. D.; Smith, R. J. Chem. Soc. Faraday Trans. I., 1978, 72, 1489.
24. Robb, I. D.; Smith, R. Polymer, 1977, 18, 500.
25. Hyde, J. S.; Dalton, L. R. in "Spin Labeling II", L. Berliner, Ed.; Academic Press: New York, 1979; Chap. 1.
26. Miller, W. G. in "Spin Labeling II", L. Berliner, Ed.; Academic Press: New York, 1979; Chap. 4.
27. Veksli, Z.; Miller, W. G. Macromolecules, 1977, 10, 686.
28. Regen, S. L. J. Am. Chem. Soc., 1974, 96, 5275.
29. Fox, T. G. Polymer, 1962, 3, 111.
30. Rudolph, W. T. "The Conformation of Synthetic Polymers Adsorbed at a Solid, Liquid Interface"; M. S. Thesis, University of Minnesota, 1976.
31. Öhrn, O. E. Arkiv Kemi, 1958, 12, 397.
32. Tuijnman, C.A.F.; Hermans, J. J. J. Polym. Sci., 1957, 25, 385.
33. Rowland, F. W.; Eirich, F. R. J. Polym. Sci. A-1, 1966, 4, 2401.
34. Mizuhara, K.; Hara, K.; Imoto, T. Kolloid Z. U. Polymere, 1969, 229, 17.

RECEIVED March 4, 1980.

2

An ESR Investigation of Environmental Effects on Nylon Fibers

K. L. DeVRIES and M. IGARASHI

College of Engineering, University of Utah, Salt Lake City, UT 84112

An important property of polymers is their ability to withstand environmental attack. While all polymers are vulnerable, some rubbers in particular are known to be susceptible to degradation agents such as ozone (1, 2, 3). Such agents are known to affect plastics but environmental studies on these materials have not received as much attention. Recently, there have been studies dealing with the degradation caused by sunlight (4, 5), oxidation (6, 7) and ultraviolet light (8, 9).

In past studies (3), Electron Spin Resonance, ESR, has been used to investigate bond rupture associated with stress-ozone degradation in unsaturated rubbers. It was believed that it might be enlightening to perform similar experiments on plastics (5, 8, 10, 11). In the study reported here the combined effects of environment (O_3, NO_2 and SO_2) and sustained stress on the ultimate properties of Nylon 6 fibers are presented. These agents were found to have a profound effect on strength, toughness and the deformation-bond rupture kinetics.

Experimental Procedures

The samples were in the form of bundles of Nylon 6 fibers (tire yarn) provided by Dr. Dusan Prevorsek of Allied Chemical Corporation. A small loading frame was constructed which would maintain an essentially constant strain on the sample over an extended length of time. The applied stress condition was monitored by a load cell. The sample was loaded in the frame with the aid of an Instron testing machine and placed in a sealed chamber in which a predetermined environment could be maintained. Figure 1 shows the equipment that maintained the strain conditions of the sample in the sealed container. The stress relaxations in some of the samples were recorded in order to know their exact stress histories. At predetermined intervals the samples were removed from the environmental chamber and unloaded. The residual strength, ultimate elongation before fracture, and the number of free radicals produced during the subsequent fracture process were determined with the aid of a servo-controlled hydraulic loading frame constructed around the magnet of a Varian E-3 ESR Spectrometer. This system (described in detail elsewhere (12)) makes it possible to simultaneously monitor and record load, deformation and free radical concentration as the sample is pulled to failure.

The sustained loading of the sample was at room temperature (23 ± 1°C) for the samples tested in Air, Ozone, SO_2, NO_2. In addition, the NO_2 tests were conducted at 81 ± 1°C in order to compare the damages between the glassy and "leathery" states. The initial loads applied to the sample during the sustained strain varied from that corresponding to a strain of 60% of what it would take to break the sample in a short time test (0.1% strain/sec) to 90% of this breaking strain. The final fracture tests on the samples, after removal from the environmental chamber, were conducted at a strain rate of approximately 0.1% strain/sec.

After these tests were conducted, a scanning electron microscope was used to examine the broken filaments and provided details about the fracture aspects of the broken fiber ends.

Results

Figure 2 shows typical ESR spectra for samples previously stressed in air, O_3, SO_2 and NO_2 environments. It is interesting that the spectrum for the sample degraded by SO_2 has a different signal as compared with the others. Analysis of the details of these spectra is continuing and will form the basis of a future paper. Figure 3 shows the effect of sustained loading in various environments on the subsequent strength of Nylon 6 samples. These environments are (a) air, (b) O_3, (c) SO_2. The breaking stress for each sample was obtained from a load to fracture test after its removal from the particular environment. Note that in this figure and all subsequent figures the samples were exposed to the environment under strains corresponding to the following fractions of the ultimate "short term breaking strain," A, 0%; B, 60%; C, 70%; D, 80%; E, 90%. In Figure 3(a) no effect was observed on strength at stresses corresponding to strains of 80% or less. Above this stress exposure for several days resulted in marked decreases both in strength and the absolute number of free radicals produced during the subsequent fracture process. Figure 3(b) shows the similar results in an ozone rich environment. Since ozone is unstable and decays to O_2 with time, these experiments were conducted with a steady flow ozone generator in the system, which produced and maintained an environmental concentration of 0.236 mole percent in the environmental chamber. Deterioration in strength was observed, with the degree of deterioration being approximately proportional both to the sustained stress and the holding time of the samples in the chamber. The samples exhibited a similar decrease in the number of bonds broken (free radicals produced) in the fracture process. In Figure 3(c) the combined effect of stress and SO_2 on deterioration in strength may

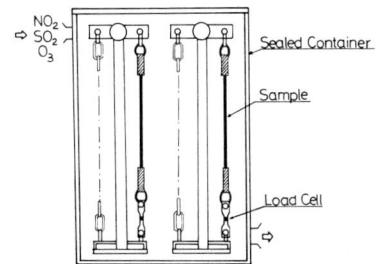

Figure 1. Loading frame which maintains constant strain on the samples

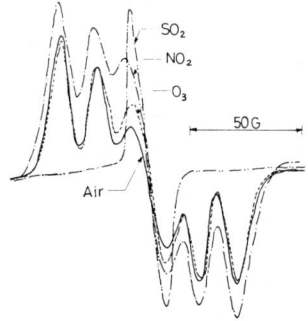

Figure 2. ESR spectra resulting from mechanical damage of Nylon 6 degraded in various atmospheres. The relative sizes of the spectra were chosen for convenience.

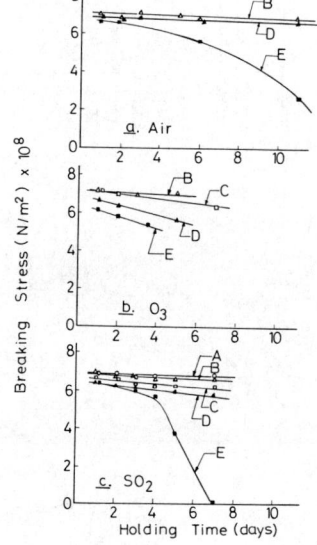

Figure 3. Degradation in strength as a function of time in various environments. All of the tests were conducted at room temperature.

be seen. The rate of deterioration is similar to, but much more rapid than that observed in air. The concentration of SO_2 was 12% by volume.

Figure 4 shows the strength of Nylon fibers stressed in various concentrations of NO_2 at 81°C for varying periods of time. Under high stress and high NO_2 concentration the degradation of the sample was almost immediate and so severe that no breaking stress is recorded in Figure 4 for this case. Figure 5 is a similar plot at the same concentration but at room temperature (23 ± 1°C). Note that the decrease in strength is much less than for the studies reported in Figures 4(b). This result is quite plausible when the mobility of the chain and its activated energy state above the glass transition temperature is considered.

Figure 6 shows the free radical concentrations produced during subsequent fracture after exposure to three different environments under strain. The net number of free radicals produced decreases almost linearly on a log-log plot in the case of NO_2, and on a semi-log plot in the cases of O_3 and SO_2. The rate of decrease of free radicals was much more rapid for those samples that were sustained at the highest strains. Note that the number of free radicals in the samples which were exposed to SO_2 environment is much lower than those samples exposed in the other atmospheres (also see Figure 2). From the short time fracture test of the Nylon 6 sample at room temperature the total number of free radicals detected was typically of the order of 10^{17}-10^{18} spins/gram. Assuming a two to one correspondence between the free radicals produced and the stress induced chain rupture, the degree of fracture or deterioration might be investigated by measuring these free radical concentrations at fracture. Figure 7 shows the relationship between the relative change in breaking stress and the decrement in the number of free radicals produced. In this figure N_o is the number of free radicals that would at zero time be obtained by extrapolation from Figure 6. σ_o is the corresponding breaking stress. This stress is essentially identical to the breaking stress obtained from short term fracture tests. N_i is the number of free radicals at fracture obtained from Figure 6. σ_i is the corresponding breaking stress obtained from Figure 3 and 4(b). Simply stated the term $(N_o - N_i)/N_o$ is the fractional decrease in the number of free radicals produced at fracture due to the combined effects of stress and environment relative to the number of free radicals produced during a standard fracture test in air. If the quantity $(N_o - N_i)/N_o$ is equal to zero, it would imply no molecular destruction due to the environmental stressing. On the other hand, if this quantity were one, it would imply destruction of all the load supporting chains prior to the final fracture tests. The fact that the curves of Figure 7 are reasonably linear might be interpreted as evidence for the hypothesis that the strength of oriented nylon is dependent on the number of and load distribution among the load carrying tie chains.

Figures 8-15 show scanning electron micrographs of the sam-

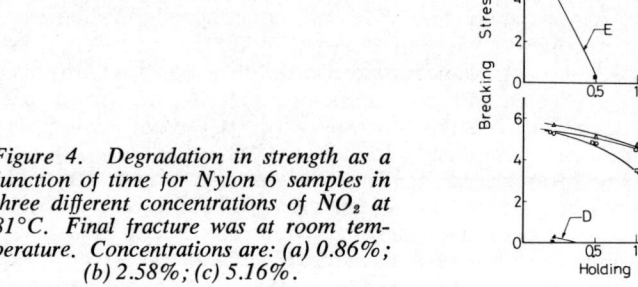

Figure 4. Degradation in strength as a function of time for Nylon 6 samples in three different concentrations of NO_2 at 81°C. Final fracture was at room temperature. Concentrations are: (a) 0.86%; (b) 2.58%; (c) 5.16%.

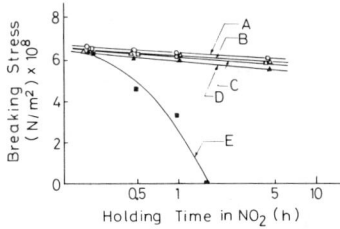

Figure 5. Degradation in strength at room temperature (23° ± 1°C) in NO_2 (conc. 2.58%) as a function of time for Nylon 6 samples

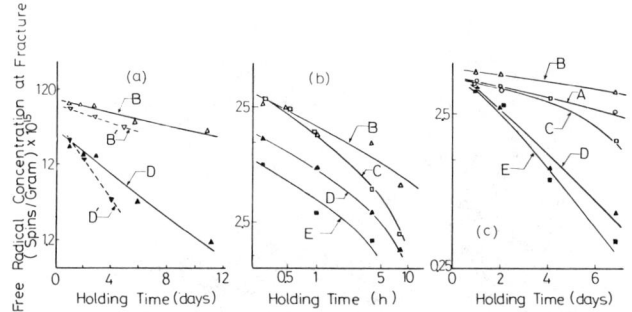

Figure 6. Free-radical concentration produced during fracture after exposure to various environments under strain as a function of holding time: (a) air (———), and O_3 (– – –); (b) NO_2 (conc. 0.86%); (c) SO_2 (conc. 12% by volume).

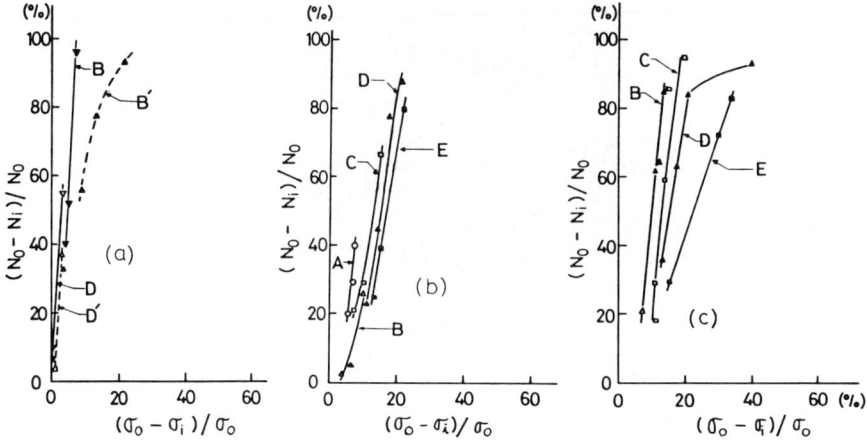

Figure 7. Observed relationship between breaking stresses and free radicals. N_0 the number of free radicals produced with no degradation; N_i, same but with degradation; σ_0, breaking strength with no degradation; σ_i, same but with degradation: (a) air (———) and O_3 (– – –); (b) SO_2 (conc. 12% by volume); (c) NO_2 (conc. 0.86%).

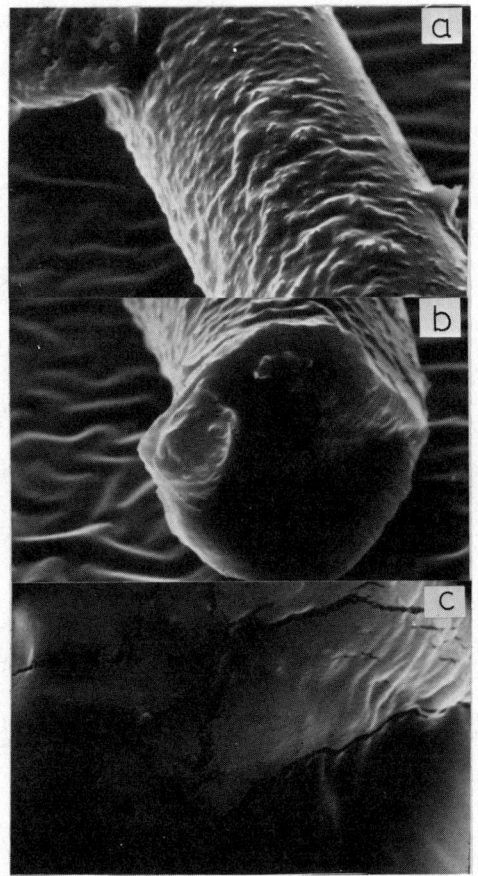

Figure 8. Fracture surface of a nylon filament exposed to an SO_2 environment. Sample was held at 90% strain for 4 d at room temperature. Magnification: (a) ca. 1370×; (b) ca. 1370×; (c) ca. 3390×.

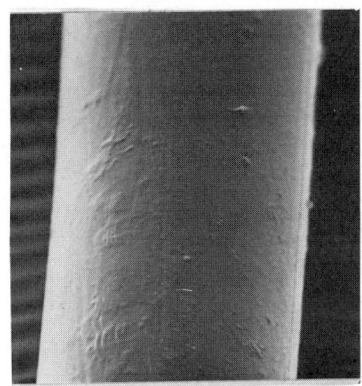

Figure 9. Nylon filament control. Surface is clean and undamaged. Magnification: ca. 1370×.

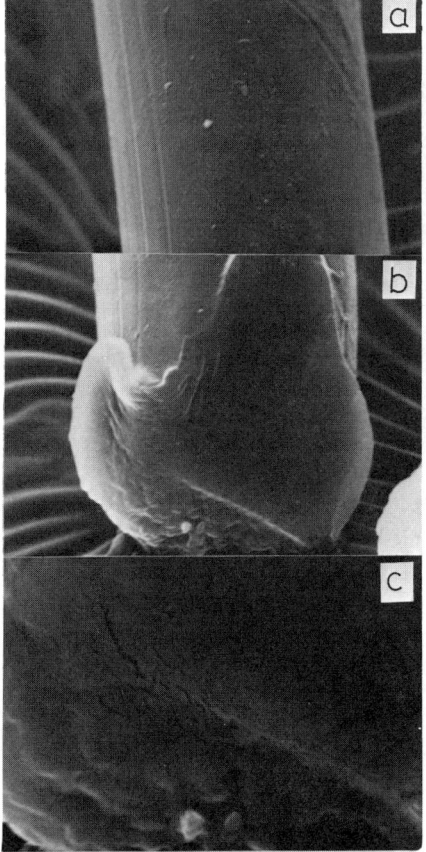

Figure 10. Fracture surface of a nylon filament exposed to an SO_2 environment. Sample was held at 70% strain for 4 days at room temperature. Magnification: (a) ca. 1430×; (b) ca. 1430×; (c) ca. 3450×.

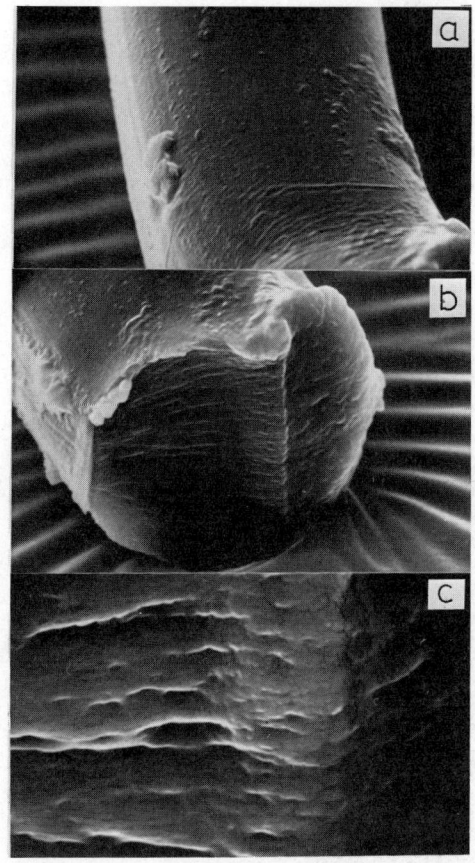

Figure 11. Fracture surface of a nylon filament exposed to an NO_2 environment at a concentration of 2.58%. Sample was held at 90% strain for 2 h at room temperature. Magnification: (a) ca. 1370×; (b) ca. 1370×; (c) ca. 3390×.

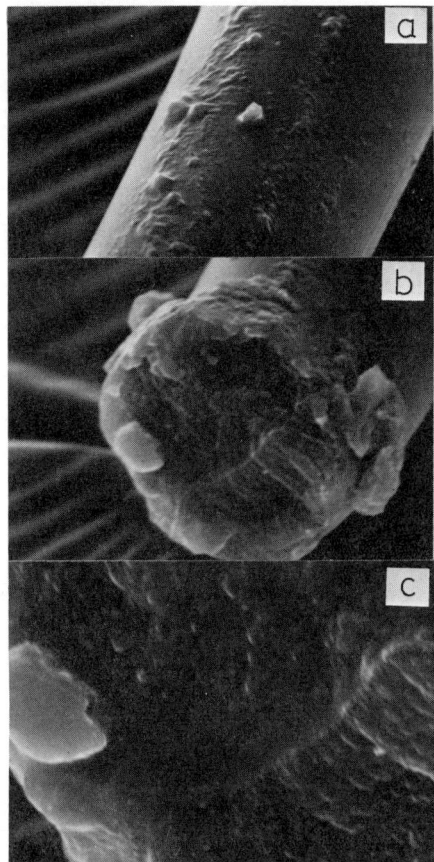

Figure 12. Fracture surface of a nylon filament exposed to an NO_2 environment at a concentration of 2.58%. Sample was held at 70% strain for 1 h at 81° ± 1°C. Magnification: (a) ca. 1550×; (b) ca. 1550×; (c) ca. 3870×.

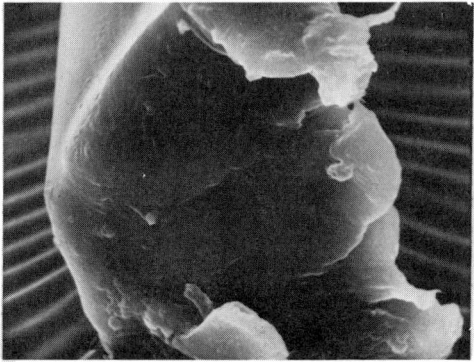

Figure 13. Surface produced by cutting with scissors: ca. 1370×.

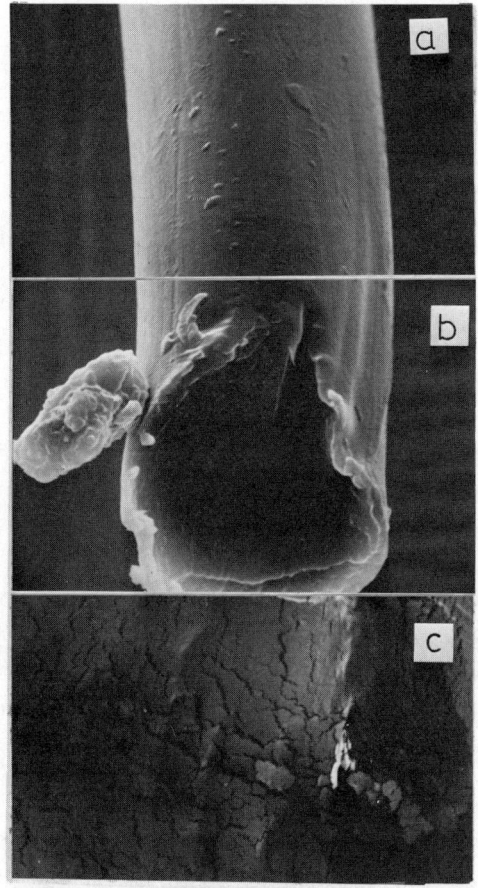

Figure 14. Fracture surface of a nylon filament exposed to an O_3 environment at a concentration of 0.1% mol. Sample was held at 90% strain for 6 d at room temperature. Magnification: (a) ca. 1370×; (b) ca. 1370×; (c) 3390×.

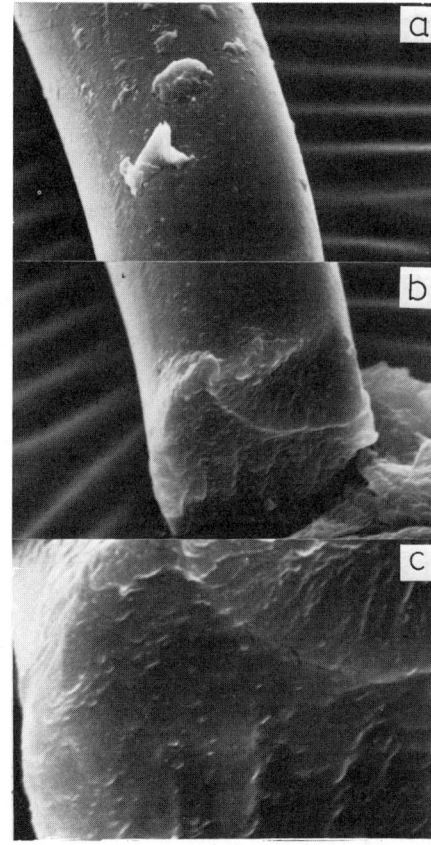

Figure 15. Fracture surface of a nylon filament exposed to an air environment. Sample was held at 90% strain for 6 d at room temperature. Magnification: (a) ca. 1250×; (b) ca. 1250×; (c) ca. 3090×.

ples. The filament in these photographs has a nominal diameter of 25 x 10^{-3} mm. In Figures 8-15 (a) shows fiber surfaces, (b) and (c) show fracture surfaces at the indicated magnification after exposure to the evironment and strains given in the captions. Figure 8 shows the fracture surface of a Nylon 6 filament which was exposed to SO_2 for four days at 90% of the maximum breaking strain and then fractured. The extent of surface degradation for this case might be illustrated by comparison with the relatively smooth surface of the standard sample which is shown in Figure 9. The SEM pictures showing the fracture aspects of these samples are similar to those described and are given in references 4 and 13. Figure 10 shows fracture surfaces which were obtained in the same enviroment as Figure 8 except at a lower sustained strain. The reader will note the marked difference in surface degradation for these two loadings, indicating a strong synergistic effect between environment and stress (strain) on surface degradation. Figures 11 and 12 show fracture surfaces of samples which were exposed to NO_2 concen-trations for similar time periods at different strain levels. In this case the perimeter surfaces are similar but there is a more pronounced difference in the fracture surfaces. The sample exposed to 90% strain (Figure 11) failed in a more "brittle" manner with much less elastic "spring back" than did the sample held at 70% strain (Figure 12). This latter failure surface is more reminiscent of samples that failed under strain without the aid of environmental attack (13). For comparison a fracture surface caused by cutting with scissors is shown in Figure 13. Figure 14 is a fracture surface produced after a sample was subjected at room temperature to an O_3 environment, at 90% strain for 6 days. The peripheral surface of the fiber does not exhibit the blistered appearance produced by SO_2 and shown in Figure 9 but does show a large number of microcracks. The microcracks on the fracture surface in these figures are in marked contrast to the comparatively "intact" fracture of a sample held under 90% strain in air and shown in Figure 15. The fracture surface of Figure 15 is nearly identical to the fracture surface of the simple tension test. J Hearle and co-workers (4, 13) studied the fracture surfaces of Nylon produced by tensile failure, and have found typical configurations of fracture surfaces. The micrographs of our Nylon samples exposed to relatively low effective agents (air, O_3) are similar to those reported by these other investigators. For the samples which were exposed to more highly degrading agents (NO_2, SO_2) the fracture surfaces were relatively flat, small cupped regions (assumed to be a crack propagating region) and also exhibited considerable microcracking.

Discussion and Conclusions

A synergistic effect between the role of stress and environmental agents such as O_3, SO_2 and NO_2 on degradation of the mechanical properties of Nylon fibers has been observed. This compliments the studies of others on degradation after light exposure (4, 5), UV light irradiation (8), and thermal-oxidation (6). It was found in the current study that the extent of degradation increased with the concentration of the oxidizing agent, the time exposure, and the applied stress (strain). Degradation in strength was accompanied by a comparable decrease in the number of bonds that had to be broken (free radicals produced) in the subsequent loading to fracture. This observation is consistent with an hypothesis that the strength of oriented Nylon is dependent on the number and orientation strength of critically loaded chains (14). Some of these would hypothetically be broken by the combined effect of the sustained stress and the chemical effects of the environment leaving fewer to sustain "stress" in subsequent loadings. One might envision thermally activated bond rupture in which the activation energy required for chain scission is aided by both the applied stress and the chemical effects of certain environmental constituents. In the subsequent fracture operation fewer of these critical chains are available to support the load, hence the reduced strength and reduction in the number of "new" free radicals produced during the subsequent loading to failure (Figure 6).

Scanning electron microscopy provides further information on the degradation caused by these enviromental agents. The extent of surface damage depended on the type and concentration of environmental agent and the magnitude of the applied stress. The fracture surface for samples degraded under stress in SO_2 and ozone exhibited extensive microcracking in comparison to samples fractured in tensile tests in air.

Literature Cited

1. Braden, M., Gent, A. N., J. Appl. Polym. Sci., 1960, 3, 100.
2. Lake, G. J., Lindley, P. B., J. Appl. Polym. Sci., 9, 1965, 2301.
3. DeVries, K. L., Simonson, E. R., Williams, M. L., J. Appl. Polym. Sci., 1970, 14, 3049.
4. Hearle, J. W. S., Lomas, B., J. Appl. Polym. Sci., 1977, 21, 1103.
5. Rosato, D. V., Schwartz, R. T., "Environmental Effects on Polymeric Materials," Vol. 2, John Willey Inc., 1968, 1306.
6. Chiang, T. C., Sibilia, J. P., J. Polym. Sci. A1, 1972, 10, 605.
7. Morand, J. L., Rubber Chem. & Tech., May-June 1972, No. 2, Vol. 50, 373.
8. Tsuji, K., Seiki, T., Polymer Letters, 1972, 10, 139.
9. Heskins, M., Reid, W. J., Pinchin, D. J., Guillet, J. E., "Photodegradation of Vinyl Chloride - Vinyl Ketone Copolymer," Chapter 19 UV-Light Induced Reactions in Polymers, S. S. Labana, Ed., No. 50, ACS Symposium Series, 1976.
10. Graves, C. T., Ormerod, M. G., Polymers, 1963, 4, 81.

11. Windle, J. J., Freedman, B., J. Appl. Polym. Sci., 1977, 21, 2225.
12. DeVries, K. L., Roylance, D. K., Williams, M. L., J. Polym. Sci., A1, 1970, 18, 237.
13. Hearle, J. W. S., Cross, P. M., J. Mater. Sci., 1970, 5, 507.
14. Lloyd, B. A., DeVries, K. L., Williams, M. L., J. Polym. Sci. A2, 1972, 10, 1415.

RECEIVED July 7, 1980.

An ESR Study of Initially Formed Intermediates in the Photodegradation of Poly(vinyl chloride)

N.-L. YANG, J. LIUTKUS,[1] and H. HAUBENSTOCK

Chemistry Department, The City University of New York, College of Staten Island, Staten Island, NY 10301

Poly(vinyl chloride), PVC, is today one of the top three most widely used thermoplastic materials. Chemically, this polymer is one of the least stable of the common polymers. The broad scope of its applications was made possible only through the development of proper technology for the processing of the polymer and the use of suitable stabilizers.

The mechanism of PVC degradation is still very poorly understood (for recent reviews, see references 1-7). Consequently, stabilizers have been developed for the polymer only in an empirical way and their performance is still far from satisfactory. Further improvements for the applications of poly(vinyl chloride) require a more detailed knowledge of the mechanism of the degradation process.

In the photodegradation process of PVC, the involvement of free radicals has been extensively documented. The applications of electron spin resonance, ESR, spectroscopy have revealed some details in the degradation mechanism (5,6). However, the potential applications of ESR spectroscopy in this area are far from being exhaustively exploited. For instance, in the initial step of the photodegradation of PVC, one would expect "Radical I" to be a predominant radical (8):

$$-CHClCH_2CHClCH_2CHCl- \xrightarrow{h\nu} -CHClCH_2\dot{C}HCH_2CHCl-$$

"Radical I"

The carbon-chlorine bond is the weakest bond in the polymer and the only one that can absorb u.v. radiation appreciably at wavelengths longer than 210 nm. "Radical I" should exhibit a six-line ESR absorption spectrum in the condensed phase due to the similar magnetic influences exerted on the paramagnetic center by the four neighboring beta-hydrogen and one

---

[1] Current address: IBM Corporation, Watson Research Center, Yorktown Heights, NY 10598.

alpha-hydrogen atoms (vide infra). However, this anticipated ESR absorption has never been experimentally established. The difficulties encountered are due to poor resolution of spectra obtained from irradiated powders or films of PVC. Dipolar broadening due to the inhomogeneous environment of the unpaired electrons in samples of poor local mobility and spin-spin interaction due to the proximity of the free radicals can lead to unresolved hyperfine absorptions. Organic glasses of polymer-solvent systems could lead to better resolved ESR spectra as demonstrated in recent radiation studies of polymers (9).

The lack of basic understanding and the prevalence of conflicting results in the area of photodegradation of poly(vinyl chloride) is mostly due to the deficiency in fundamental information about the sequence of events that takes place after the polymer molecule absorbs a photon. The use of PVC samples containing foreign unknown substances and the attribution of anomalies and discrepancies in experimental results to impurities further created considerable confusion in this subject. At the present stage of development in the field of photodegradation of PVC, it is critical to establish for pure PVC the identities and dynamics of the intermediates involved.

We report here well-resolved ESR spectra obtained from amorphous solid-solution, i.e. glass, PVC-solvent systems, and further demonstrate that the predicted six-line spectrum of "Radical I" at the initial stage of photodegradation can be obtained at liquid nitrogen temperature, LNT. ESR spectra in glass systems for the photodegraded copolymer of vinyl halides and low molecular weight alkyl chlorides were also obtained in order to establish the identities of radicals involved in PVC photodegradation.

Sample Preparation

Polymer and copolymer.
Polymer and copolymer were prepared in our laboratory for strict control of purity, branching content and molecular weight. All monomers and solvents were carefully purified. PVC samples were prepared as powders in bulk using free radical photoinitiation with either AIBN or uranyl nitrate as initiators. The AIBN/UV initiation was used for photopolymerization at temperatures of 50°, 25°, and 0°C. A uranyl nitrate/absolute ethanol initiation system with visible light source, 500W tungsten lamp, was used for polymerization at -78°C (10). This system was also used at higher polymerization temperatures for comparison with PVC samples prepared by AIBN/UV initiation. Although it had been suggested (10) that the use of UV initiation might cause degradation during polymerization, no such evidence was found. The pyrex glass apparatus apparently was not transparent to U.V. of short wavelength.

The polymerization systems were always deoxygenated thoroughly by freezing at liquid nitrogen temperature and thawing and evacuated under high vacuum. The polymer samples produced were washed exhaustively with absolute ethanol and distilled water. The most important variables were polymerization temperature and % conversion. Chain transfer to monomer is the controlling factor in determining the molecular weight of the polymer. For PVC, a decrease in irradiation time will cause a decrease in % conversion. This will minimize branching since the branching mechanism involves chain transfer to polymer (10,11). Conversions of less than 10% were shown to preclude branching in bulk polymerized vinyl chloride at -10° to 30°C (12). At low conversion decreasing the polymerization temperature will decrease branching (11,13), increase molecular weight (14), and increase crystallinity (15,16).

Copolymers of vinyl chloride-vinyl bromide were prepared in bulk using the same conditions in the preparation of PVC, except that only uranyl nitrate was used as in initiator with visible light to avoid photodegradation during polymerization. Thermal analysis of the copolymer indicated random placement of the two comonomers.

Alkyl Chlorides. 3-Chloropentane was prepared by the reaction of thionyl chloride with 3-pentanol. Hexamethyl-phosphoric triamide, HMPT, was used as a complexing agent to ensure substitution at the 3 position (17). The organic layer was dried with magnesium sulfate and distilled. The middle fraction, b.p. 104°C, was collected for use. 1-Chloropentane was purified by repeated distillations. The above preparation procedures ensure maximum purity of the compounds for ESR study.

Glasses. The solvents used to form the glass with polymers should be clear in the UV region to allow maximum absorption of radiation by the polymer. The ESR of irradiated solvents should not interfere with that of the polymer. Three solvents were found to be close to ideal: tetrahydrofuran (THF); p-dioxane (DX); and tetrahydropyran (TP). All three were purified by repeated distillation and column chromatography. Poly(vinyl chloride) and the copolymer solutions (5-15%, w/v) were prepared from these solvents, degassed, sealed under vacuum (10^{-6}:torr), and irradiated.

ESR Spectra

At liquid nitrogen temperature, the ESR spectra of PVC and copolymer, COP, glasses were identical. Varying the solvent only affected the spectra slightly. The glasses with THF, DX, and TP appear as similar six line absorptions in the ESR (Figure 1). The spectra are all symmetrical, with a hyperfine splitting of 23 G, line intensity ratios of 1:4:9:9:4:1, and g-value of 2.0036 (reference DPPH). This well-resolved ESR spectrum has never been observed before in the photodegradation of PVC. It should be noted here that in the radiolysis of PVC, a sequence of radicals has been monitored by an optical absorption method (18,19).

The glasses were colorless before and after irradiation at LNT, indicating the absence of polyenyl structures due to dehydrochlorination. On warming to room temperature, RT, the PVC glasses became light yellow, and the COP glasses a light orange. These color changes indicate that the radical formed at LNT is a precursor to the polyene radical.

The low molecular weight analogs, 3-chloropentane (3-CP) and 1-chloropentane (1-CP) were treated in exactly the same manner as the polymer glasses. Since these analogs were small molecules, no solvent was necessary to enhance resolution in the ESR. On irradiation, cleavage of the carbon-chlorine bond is expected for both. The 3-CP should serve as a model to indicate carbon-chlorine cleavage in the main PVC chain. The 1-CP molecule could be used to indicate formation of any primary radicals in PVC from branch points, chain cleavage, etc.

The ESR spectrum of 3-CP irradiated and measured at LNT is shown in Figure 2. The spectrum is almost identical to that of the PVC/TP glass with the same hyperfine splitting, intensity ratio, and g-value. This result should not be surprising, since carbon-chlorine cleavage was expected to be the major process in each, and both molecules can be considered as $R-CH_2-CHCl-CH_2-R$. A one-to-one solution of 3-CP in tetrahydropyran was prepared and irradiated under identical conditions. The spectrum was essentially the same.

The 1-CP analog, without solvent, also gave rise to a six line absorption (Figure 3). The hyperfine splitting and g-value are similar to that for 3-CP, but the line intensity ratio is quite different, 1:4:4:4:4:1. This ratio is indicative of primary radicals. Since this type of intensity ratio was not observed for the polymer glasses, no significant contribution from primary radicals is likely.

Discussion

"Radical I". The purpose of polymerizing vinyl chloride under a variety of conditions was to prepare polymers with a wide range of branching, and molecular weight. The ESR spectra of all PVC glasses, however, exhibited the same six line absorption at LNT. This seems to indicate that the variation of the above properties in the present case did not affect the degradation process of PVC appreciably.

Branching should incorporate weaker tertiary bonds in PVC, either carbon-hydrogen or carbon-chlorine. Secondary and tertiary C-H bonds have bond dissociation energies of 94 and 90 Kcal/mole respectively. A tertiary carbon-chlorine bond may require even more energy for homolytic cleavage than a secondary one, i.e. 75 versus 73 Kcal/mole (20). Cleavage of the branch

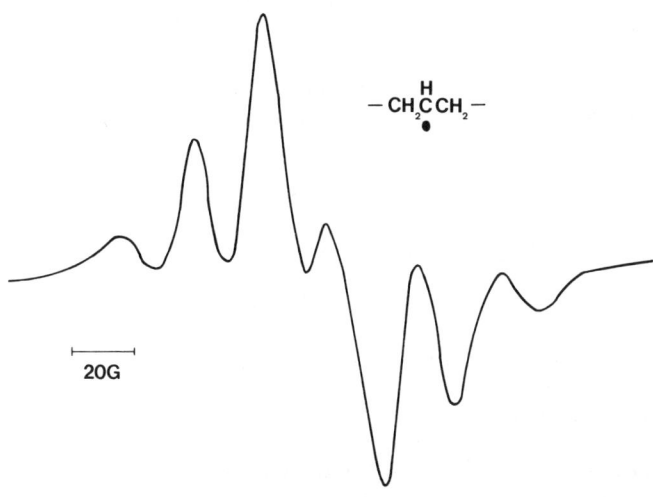

Figure 1. ESR spectrum of UV-irradiated PVC–TP glass at liquid nitrogen temperature

Figure 2. ESR spectrum of UV-irradiated 3-chloropentane at liquid nitrogen temperature

itself is improbable due to the bulky nature of the group. If
cleaved at LNT, this group would be expected to rapidly recombine
due to cage effect. In addition, the evidence indicates that one
radical species predominates in the ESR spectrum. Therefore, the
effect of branching may not be important in providing initial
sites for degradation in the present case.

Poly(vinyl chloride) prepared at -78°C, and containing no
branches, was compared to PVC prepared at 50°C and having the
highest branch content. Similar comparisons were made for
polymers prepared with intermediate branch content. No
differences were observed in the ESR spectra on irradiation and
measurement at LNT. Even for the 50°C case the number of branches
are only four per thousand repeat units (13). Therefore, even if
radicals are formed at (or near) these branch sites, their effects
will be overshadowed by the unbranched radicals if they do not
significantly alter the degradation mechanism.

The molecular weight (M_V) of PVC prepared at 50°C was found
to be 31,000 from intrinsic viscosity measurements. PVC
prepared at -78°C is expected to have the highest molecular
weight, ca. 100,000. Aside from crystallinity and branch content, these polymers should differ only in number of end groups
per unit weight of sample. Disproportionation reactions often
terminate a growing PVC chain (17), resulting in unsaturated
chain ends. These chain ends may be able to initiate degradation
by providing labile allylic atoms for cleavage by UV. The net
effect of variable molecular weights is to provide correspond-
ingly variable numbers of end groups. In the absence of
quantitative photo-chemical measurement, such effects can not
as yet be ascertained. Of special interest was the possible
detection of radicals from branch points or chain ends. Attempts
to detect any other ESR absorption preceding the sextet observed
at LNT were not successful.

Two types of initiator were used in this work, AIBN and
uranyl nitrate. Unreacted initiator was thoroughly removed from
the polymer. A comparison of PVC samples prepared by the two
different initiators showed identical behavior under ESR
investigation. Therefore, end groups derived from the
initiator do not appear to affect the course of degradation
appreciably.

The vinyl chloride and vinyl bromide copolymer samples were
prepared as polymer analogs of PVC. Cleavage of the carbon-
halogen bond at LNT should result in the same radical for these
polymers. Since secondary C-Cl and C-Br bond dissociation
energies are 73 and 59 Kcal/mole respectively, a copolymer of
vinyl chloride and vinyl bromide could be regarded as a PVC chain
with weak points. The feed ratio of VC/VB was 10/1 by volume.
Since the reactivity ratios are (in solution, at 40°C),
r_1 = 0.825 for VC and r_2 = 1.050 for VB (22), the copolymer
composition should be 15 units of vinyl chloride for one unit of
vinyl bromide on the average. In addition, the values for r_1 and

r_2 are close to unity, indicating formation of a random copolymer. Preferential scission is expected at the C-Br bond and the resulting polymer radical would be identical to that formed by C-Cl cleavage in PVC. Hence all the COP glasses exhibited ESR spectra identical to the PVC glasses at LNT.

Since identical ESR spectra were obtained on U.V. irradiation at liquid nitrogen temperature of PVC, copolymer of vinyl chloride with vinyl bromide and 3-chloropentane it is concluded that this radical is the "Radical I." This radical may play an important role in PVC degradation. To further firmly establish the identity of this radical, theoretical analysis of "Radical I" and computer simulations of ESR spectrum were performed. The polymer radical ESR spectra are very sensitive to the conformation of the chain. The characteristic chain conformation of vinyl syndiotactic sequences in solution has been shown (23) to consist chiefly of trans-trans groups, separated by gauche units: ... $(TT)_x (GG)_1 (TT)_y (GG)_1 (TT)_z$ ... . Crescenzi (24) proposed a similar structure: (TT) (TT) (GG) (TT) (TT) ... (TT) (GG) (TT) ... for syndiotactic chains in solution. In addition, Corradini (24) calculated that the conformational internal energy of syndiotactic PVC was at a minimum for a (TTTT) sequence.

Since the prepared PVC samples had a high degree of syndiotacticity, it is expected that the trans planar grouping should predominate for PVC in solution. On irradiation, the radical would reside most likely in one of the trans-trans sequences. If the proposed radical, $-CH_2-\overset{\bullet}{CH}-CH_2-$ is formed, it follows that the four beta protons must be equivalent from geometric considerations (Figure 4). There are only two dihedral angles that fulfill this requirement, $\theta = 60°$ or $30°$. A dihedral angle of $60°$ would not be in accord with the trans-planar sequence. Therefore, the dihedral angle must be $30°$.

This angle should be used in the McConnell relationship, along with appropriate values for A and B. The spin density on a saturated secondary carbon, the isopropyl radical, was reported to be 0.844 (25), which was used in the analysis of the analogous PVC radical. Thus, the hyperfine splitting of the beta protons was calculated to be 25.3 G. The equation $\Delta H = Q\rho$, determined the amount of spin polarization for the alpha hydrogen. The values $Q = 26.2$ and $\rho = 0.844$ (for secondary alkyl radicals) gave 22G as the hyperfine splitting from the alpha hydrogen.

Therefore, due to the four equivalent beta protons, a quintet is expected, which should be further split into ten lines by the alpha proton (Figure 5). This constructed spectrum can be applicable to either 3-chloropentane or PVC, the 'R' groups being CH_3 or polymer chains respectively. If this PVC radical could be observed in the liquid state, all ten lines would be visible. However, for the solid at LNT, the central eight lines will overlap, and the spectrum will appear as a symmetrical sextet with a theoretical line intensity ratio of 1:5:10:10:5:1.

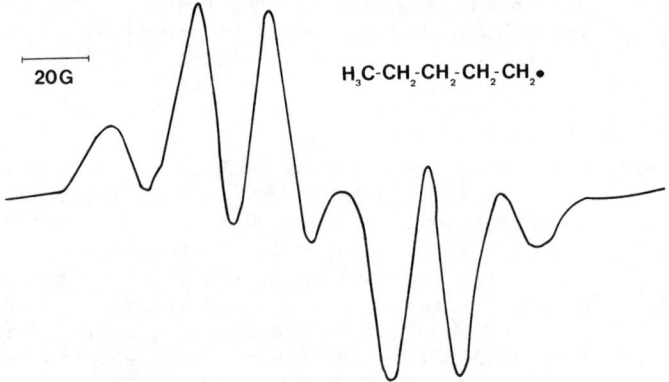

Figure 3. ESR spectrum of UV-irradiated 1-chloropentane at liquid nitrogen temperature

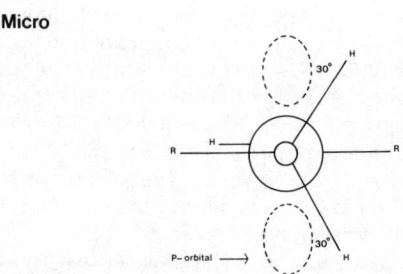

$\Delta H_a = Q\rho = (26.2)(0.844) = 22G$

$\Delta H_\beta = (A + B\cos^2\theta)\rho = (0 + 40\cos^2 30°) \, 0.844 = 25.3G$

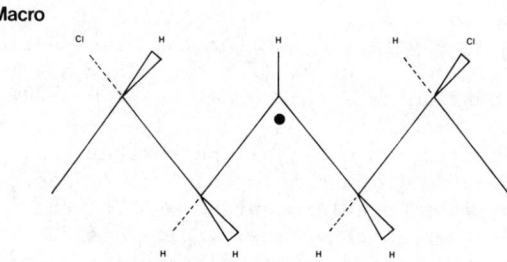

Figure 4. Conformation of PVC radical, i.e. "Radical I"

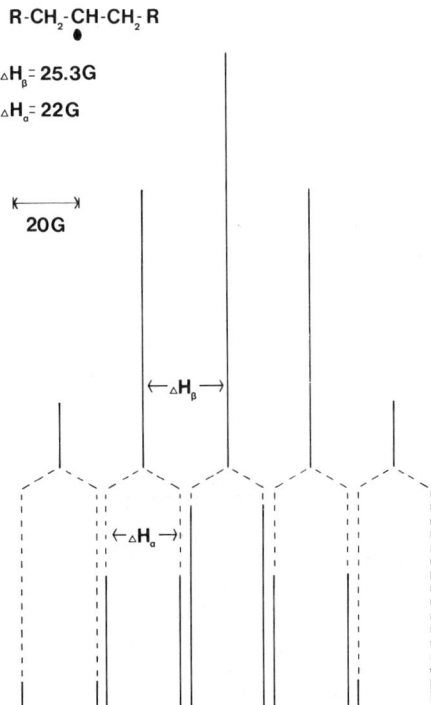

Figure 5. ESR stick spectrum of PVC radical at liquid nitrogen temperature

A computer simulation was performed to observe the effect of variable line width on the calculated ESR first derivative spectrum. Since dipolar interaction is the major contribution to line broadening in the ESR spectrum of PVC radicals, a Gaussian line shape is expected for each of the ten absorptions. Therefore each line was assigned a Gaussian shape, the variables being: relative amplitude; position in the spectrum; and line width. The resultant curves were then added, and the first derivative taken.

The equation for a first derivative Gaussian curve is (26):

$$Y'(H) = \frac{-2y_m \ln 2}{\frac{1}{2}\Delta H_{\frac{1}{2}}} \left(\frac{H - H_o}{\frac{1}{2}\Delta H_{\frac{1}{2}}}\right) \exp\left[-\left(\frac{H - H_o}{\frac{1}{2}\Delta H_{\frac{1}{2}}}\right)^2 \ln 2\right]$$

where: y_m is the relative amplitude of each line; $\Delta H_{1/2}$ is the line width at half the maximum amplitude; H_o is the reference position (origin); and H is the position of the absorption. The relative amplitudes and line positions were as shown in Figure 5. This equation was normalized and transformed into a Fortran statement for an IBM 370/168 computer. The computed spectra were displayed graphically by an IBM-1625 Plotter. Figure 6 represents the simulated case for PVC glasses at LNT. Broadening effects have transformed the observed spectrum to a sextet with no apparent fine structure. The computed apparent hyperfine splitting is 23.8 G and the line intensity ratio is 1:5:10:10:5:1. These results offer further evidence for the proposed 3-CP and PVC radicals, and support the contention that the observed sextet is actually an unresolved ten line absorption.

"Radical II," a Precursor for the Polyenyl Radical. The change of the ESR spectrum of irradiated PVC with temperature is very revealing. When the PVC/THF glass was warmed to -110°C after U.V. irradiation at LNT, the six line spectrum was replaced by a five line absorption (Figure 7). Under the same conditions, tetrahydrofuran alone does not exhibit such an absorption. The ESR absorption from the PVC/THF glass had a hyperfine splitting of 19-21 G, and g-value of 2.0036. The line intensity ratio appears to increase as the center of the spectrum is approached (theoretical value 1:4:6:4:1), but interference from a central singlet, probably solvent radicals, did not permit quantification. This spectrum has not been reported for the degradation of PVC. It was determined that the spectral change of sextet at LNT (-196°C) to quintet at -110°C was not a conformational conversion, since on cooling from -110°C to LNT, the same quintet was observed. Therefore, a radical conversion must have occurred.

With reference to the parent polymer molecule, the most likely assignment for the radical giving rise to the observed spectrum is:

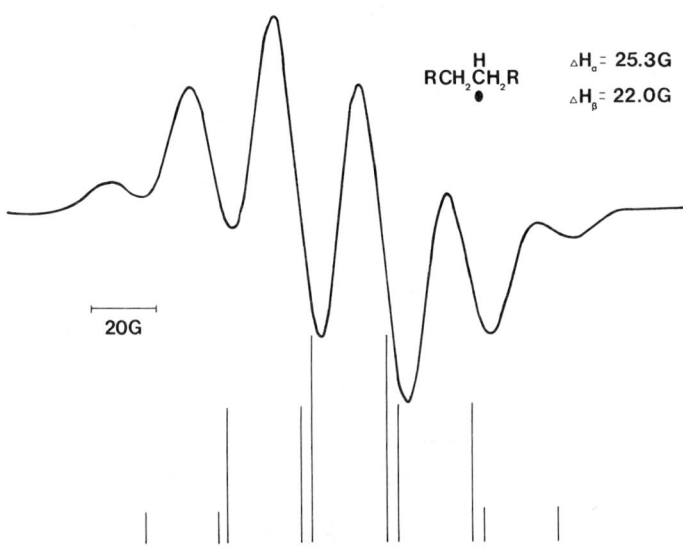

Figure 6. Computer-simulated ESR spectrum of PVC radical with $\Delta H_{1/2} = 24$ G

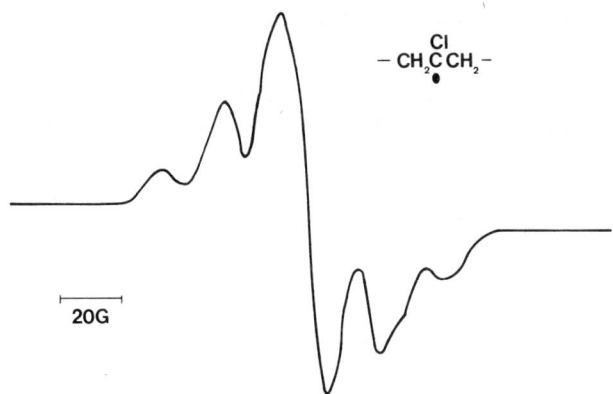

Figure 7. ESR spectrum of PVC–THF at $-110°C$, UV irradiated at liquid nitrogen temperature

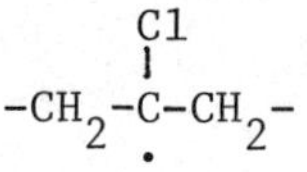

$$-CH_2-\overset{\overset{\displaystyle Cl}{|}}{\underset{\bullet}{C}}-CH_2-$$

"Radical II"

This radical would produce a quintet if the beta hydrogens were equivalent. In polymer radicals of the type $-CH_2-\overset{\bullet}{C}X-CH_2-$, the methylene protons were shown to be equivalent (25). In addition, the contribution of an alpha chlorine to hyperfine splitting is listed as 2.8 G for aliphatic radicals (27). Since this splitting is too small to be observed in the spectrum, a quintet would be expected from the four equivalent beta protons. The dihedral angle could not be accurately calculated for this radical, since information on the spin density of the alpha carbon is unavailable. It is expected that the spin density in this case should be less than that for "Radical I," since delocalization of the unpaired electron onto chlorine is possible. As the electronegativity of the substituent (X) on the alpha carbon increased, the spin density decreased (27). This lower spin density in "Radical II" should produce a decrease in hyperfine splitting (as predicted by the McConnell relationship). This decrease was observed; for Radical I, the sextet is characterized by $\Delta H = 23$ G, while Radical II has a $\Delta H = 20$ G. Therefore it is possible that the dihedral angle is the same for both radicals, even though the observed splitting is different.

It is noteworthy that "Radical II" is more stable than "Radical I." It was found that radicals of the type R-CCl-R were less reactive than the corresponding R-$\overset{\bullet}{C}$H-R radical (28). This observation is in keeping with our ESR results.

Conclusion

In summary, our study of the photodegradation of poly(vinyl chloride) have established the following radical intermediates:

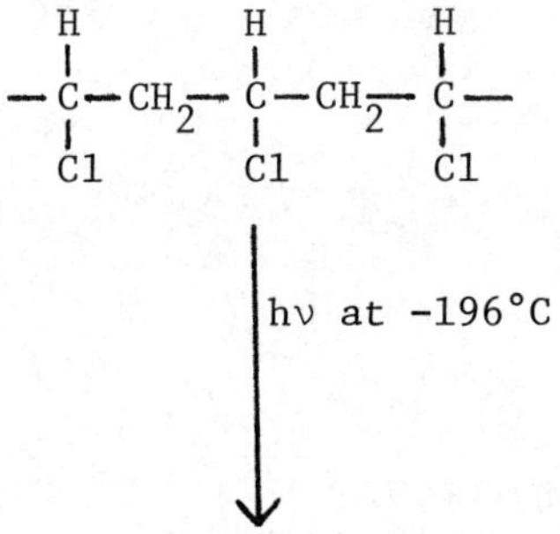

$$-\overset{\overset{\displaystyle H}{|}}{\underset{\underset{\displaystyle Cl}{|}}{C}}-CH_2-\overset{\overset{\displaystyle H}{|}}{\underset{\underset{\displaystyle Cl}{|}}{C}}-CH_2-\overset{\overset{\displaystyle H}{|}}{\underset{\underset{\displaystyle Cl}{|}}{C}}-$$

$\Big\downarrow$ hν at $-196°C$

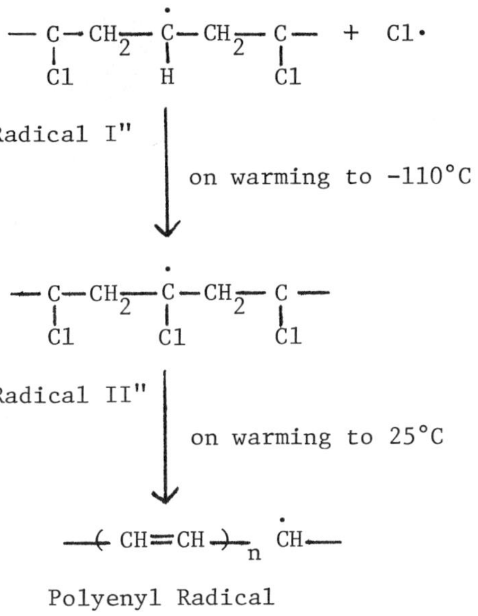

$$-\overset{\bullet}{\underset{\underset{Cl}{|}}{C}}-CH_2-\underset{\underset{H}{|}}{C}-CH_2-\underset{\underset{Cl}{|}}{C}- \quad + \quad Cl\cdot$$

"Radical I" ↓ on warming to −110°C

$$-\underset{\underset{Cl}{|}}{C}-CH_2-\overset{\bullet}{\underset{\underset{Cl}{|}}{C}}-CH_2-\underset{\underset{Cl}{|}}{C}-$$

"Radical II" ↓ on warming to 25°C

$$-(CH=CH)_n-\overset{\bullet}{C}H-$$

Polyenyl Radical

The chlorine radicals, $^{35}Cl\cdot$ and $^{37}Cl\cdot$, could not be detected in the present study due to its short spin relaxation time even at the low temperature of liquid nitrogen (29). The conversion of "Radical I" to "Radical II" was probably via hydrogen abstraction.

Mechanisms of photodegradation of PVC offered in the literature frequently cite the polyene radical with little experimental evidence for its precursors. In the present work, with the application of the PVC-glass systems, we have established "Radical I" and "Radical II" as intermediates. ESR results in the PVC-solvent glasses indicate a promising route for further fundamental investigation. Quantitative investigations of photochemical generation of radicals and kinetics of radical processes are in progress.

Acknowledgement

This work was supported by grants (No. 12292 and 13124) to N.L.Y. through the PSC-BHE Award Program. This paper is based on J. Liutkus' doctoral dissertation submitted in partial fulfillment of the requirements for the degree of Doctor of Philosophy at the Graduate School, C.U.N.Y.

Literature Cited

1. Braun, D. in "Degradation and Stabilization of Polymers," Genskens, G., Ed.; Halsted Press, N.Y., N.Y., 1975, p. 23.
2. Ranby, B.; Rabek, J.F.; Canback, G. J. Macromol. Sci.-Chem., 1978, A12, 587.

3. Owen, E.D. in "Ultraviolet Light Induced Reactions in Polymers," (ACS Symposium Ser., No. 25), Labana, S.S., Ed.; American Chemical Society: Washington, D.C., 1976.
4. Close, L.; Gilbert, R. Polym. Technol. Eng., 1977, 8, 177.
5. Ranby, B.; Rabek, J.F. "Photodegradation, Photo-oxidation and Photostabilization of Polymers": Wiley, New York, 1975; p. 192.
6. Ranby, B.; Rabek, J.F. "ESR Spectroscopy in Polymer Research": Springer-Verlag, New York, 1977; p. 209.
7. Salovey, R. in "The Radiation Chemistry of Macromolecules, Vol. II," Dole, M., Ed.; Academic Press: New York, N.Y., 1973; p. 37.
8. Lawton, E.J.; Balwit, J.S.; Powell, R.S. J. Chem. Phys., 1960, 33, 395.
9. Chung, Y.J.; Yamakawa, S.; Stannett, V. Macromolecules, 1974, 7, 204.
10. Manson, J.A.; Iobst, S.A.; Acosta, R. J. Poly. Sci., 1972, A1, 10, 179.
11. Boccato, G.; Rigo, A.; Talamini, G.; Grandi, F. Mak. Chem., 1967, 108, 218.
12. Madruga, E.L.; Millan, J. J. Poly. Sci., 1974, A1, 12, 2111.
13. Nakajima, A.; Hamada, H.; Hayashi, S. Mak. Chem., 1966, 95, 40.
14. Pham, Q.; Millan, J.; Madruga, E. Mak. Chem., 1974, 175, 945.
15. Hassan, A.M. J. Poly. Sci., 1974, A2, 12, 655.
16. Nakajima, A.; Kato, K. Mak. Chem., 1966, 95, 52.
17. Hudson, H.; Spinoza, G. J. Chem. Soc., 1976, P1, 109.
18. Salovey, R.; Luongo, J.P.; Yager, W.A. Macromolecules, 1969, 2, 198.
19. Salovey, R.; Albarino, R.V.; Luongo, J.P. Macromolecules, 1970, 3, 314.
20. Steacie, E.W.R. "Atomic and Free Radical Reactions," Vol. 1, 2nd Ed., Reinhold, N.Y., N.Y., 1954; p. 97.
21. Koleske, J.V.; Wartman, L.H. "Polyvinylchloride," Gordon & Breach, N.Y., 1969.
22. Michael, A.; Schmidt, G.; Guyot, A. J. Macromol. Sci.-Chem., 1973, A7(6), 1279.
23. Morawetz, H. "Macromolecules in Solution," Interscience, N.Y., 1965.
24. Ketley, A.D. "The Stereochemistry of Macromolecules," Marcel Dekker, Inc., N.Y., 1968.
25. Butyagin, P.; Dubinskaya, A.M.; Radstig, V.A. Russ. Chem. Rev., 1969, 38(4), 290.
26. Poole, C.P., Jr. "Electron Spin Resonance," Interscience: N.Y., N.Y., 1967; p. 797.
27. Fisher, H. Z. Naturforsch., 1965, 20a, 428.
28. Kharasch, M.S. J. Amer. Chem. Soc., 1951, 73, 632.
29. Vanderkoui, N., Jr.; MacKenzie, J.S. "Free Radicals in Inorganic Chemistry," Adv. Chem. Series, 1962, 36, 98.

RECEIVED March 4, 1980.

4

Manganese(II) and Gadolinium(III) EPR Studies of Membrane-Bound ATPases

CHARLES M. GRISHAM

Department of Chemistry, University of Virginia, Charlottesville, VA 22901

Membrane enzymes are involved in many essential cellular functions, including transport of substances in and out of cells or organelles, energy transduction, hormonal control, lipid metabolism and cellular control (1,2,3,4). In recent years, the development of isolation procedures for membrane enzymes (5-10) has proceeded at such a rapid pace that the use of certain physical probe methods for the determination of structure and function in these systems has begun to lag behind the isolation procedures. For example, paramagnetic probes have been used in a large variety of ways to characterize the structure and mechanism of action of numerous soluble enzymes (11,12,13,14,15), but have not been as widely used with membrane enzymes. Our own interests in cation transport and in the ATPase enzymes which are often responsible for this ion transport have led us to a characterization of three membrane-bound ATPases from mammalian systems (9,16-24). These enzymes are the ($Na^+ + K^+$)-ATPase, a plasma membrane enzyme responsible for Na^+ and K^+ transport in mammals, the Ca^{2+}-ATPase, which is responsible for Ca^{2+} uptake in sarcoplasmic reticulum, and a newly purified Mg^{2+}-ATPase, whose transport function, if any, is unknown. Characterization of the structure and function of these membrane enzyme systems will depend on the development of specific spectroscopic probes. In this regard, use can be made of the monovalent and divalent cation requirements of the system as well as the substrate and inhibitor binding sites. Thus our previous studies have partially characterized a Mn^{2+} binding site on the ($Na^+ + K^+$)-ATPase which is responsible for the divalent cation activation of the enzyme and the transport system (16,17,18). NMR studies using the $^{205}Tl^+$, ^{31}P, and $^7Li^+$ nuclei have also shown that this Mn^{2+} site is very near (a) a Na^+ site which is probably involved in enzyme activation and ion transport (17), (b) a K^+ site which has not previously been detected in kinetic studies (19,20), and (c) a noncovalently binding phosphate site (18) which had not been detected previously on this enzyme.

0-8412-0594-9/80/47-142-049$08.00/0
© 1980 American Chemical Society

Electron paramagnetic resonance (EPR) spectra of bound metal ions can provide additional insight into molecular motion and structure at the active site of enzyme-metal complexes. Magnetic anisotropies which arise from asymmetry in the electronic environment of the ion are often apparent in the spectra, and changes in these parameters can reflect changes in the ligand composition and geometry of the complex. Correlation of parameters from the EPR spectra with data obtained from NMR studies of ligands, substrates and metal-bound water molecules can provide unique information on the catalytic centers of enzymes.

The present paper will outline some of our recent EPR and NMR studies using Mn^{2+} as a paramagnetic probe of sheep kidney $(Na^+ + K^+)$-ATPase and Gd^{3+} as a paramagnetic probe of sarcoplasmic reticulum Ca^{2+}-ATPase. Estimates of the relevant electron spin relaxation times and some features of the interaction between substrates and activators with the enzyme-metal complexes will be inferred from the EPR spectra and the accompanying nuclear relaxation data.

Theoretical Basis for Mn(II) and Gd(III) EPR Studies of Membrane Enzymes

A. Mn(II) EPR. The five unpaired 3d electrons and the relatively long electron spin relaxation time of the divalent manganese ion result in readily observable EPR spectra for Mn^{2+} solutions at room temperature. The Mn^{2+} ($S = 5/2$) ion exhibits six possible spin-energy levels when placed in an external magnetic field. These six levels correspond to the six values of the electron spin quantum number, M_S, which has the values 5/2, 3/2, 1/2, -1/2, -3/2 and -5/2. The manganese nucleus has a nuclear spin quantum number of 5/2, which splits each electronic fine structure transition into six components. Under these conditions the selection rules for allowed EPR transitions are $\Delta M_S = \pm 1$, $\Delta m_I = 0$ (where M_S and m_I are the electron and nuclear spin quantum numbers) resulting in 30 allowed transitions. The spin Hamiltonian describing such a system is

$$\mathcal{H} = gBH \cdot S + hAS \cdot I + D[S_z^2 - (1/3)S(S+1)] + E[S_x^2 - S_y^2]$$

where S and I are the electron and nuclear spin operators, g and A are the Zeeman and hyperfine interaction constants, and D and E are the axial and rhombic distortion parameters of the zero field splitting (ZFS) interactions (25). The ZFS interactions are highly anistropic and the effect of these terms on the energy level is orientation dependent. In rapidly rotating complexes of manganese (II) ions the terms involving D and E are effectively averaged to zero, since $\Delta\omega\tau_r < 1$. Here τ_r is the rotational correlation time of the complex and $\Delta\omega$ is the ZFS in rad/sec. Since ZFS terms make no net contribution to the spacing of energy levels, transition frequencies for all of the electronic fine structure components are equal and the 30 allowed transitions thus give rise to six inhomogeneously broadened lines corresponding to the six values of m_I. Within each of the six resolved lines are the five unresolved fine structure components.

On the other hand, when rotational motion is slow, or when the symmetry of the complex is less than cubic, as in Mn^{2+} complexes with macromolecules, $\Delta\omega\tau_r$ is often greater than one, the anisotropic interactions are incompletely averaged and EPR spectra similar to those for randomly oriented solid samples are observed. In these cases the spectra depend upon the angular relationships between the magnetic field vector and the crystal field axis of the ion. Moreover, when the symmetry of the manganese ion complexes deviate greatly from cubic, the EPR spectra depend upon the sharing of spectral intensity between the normal and forbidden ($\Delta M_s = \pm 1$, $\Delta m_I = \pm 1$) transitions.

The impact of solvent molecules and the resulting transient distortions of the Mn^{2+} complex determine the electron spin relaxation time of the system (26). Thus efficient solvent collisions at the bound Mn^{2+} will yield broad EPR lines, while narrow lines should result when Mn^{2+} is inaccessible to this rapid, fluctuating motion.

B. Gd(III) EPR. The literature on Gd(III) EPR in solution is not extensive. Hudson and Lewis (27) have presented a theory for the electron spin relaxation of ^8S ions (e.g., gadolinium(III)) in solution. These authors assumed that the dominant line broadening mechanism for an ^8S ion is provided by the modulation of the zero-field splitting by a process with a characteristic time τ. The transverse relaxation rate is given by

$$\frac{1}{T_{2e}} = -ZM(\omega,\tau)$$

where Z is the inner product of the zero-field splitting tensor (in $rad^2 sec^{-2}$) and $M(\omega,\tau)$ is the relaxation matrix, ω being the electron Larmor frequency. Z will vary with the strength and symmetry of the ligand field and will express the dynamic process responsible for relaxation. This process could be either the rotation of the complex (this would essentially be rotational modulation of a static zero-field splitting) or symmetry fluctuations in the complex due to distortions induced by the impact of solvent molecules (i.e., modulation of the transient zero-field splitting).

Mn(II) EPR Studies of Sheep Kidney $(Na^+ + K^+)$-ATPase and Mg^{2+}-ATPase

A. Binding Studies. The ability to use Mn^{2+} EPR spectra to quantitate Mn^{2+} binding to a macromolecule derives from the observation made 25 years ago by Cohn and Townsend (28) that the intensity of the Mn^{2+} EPR spectrum is proportional to the concentration of free manganous ion. When Mn^{2+} forms a complex, the X-band EPR spectrum is broadened and its intensity is greatly diminished relative to that of the free ion. The data can be plotted in the form of a Scatchard plot, but care should be taken to observe the limits of free and bound ligand (Mn^{2+}) suggested by Deranleau (29). In a theoretical consideration of the Scatchard plot as a tool for ligand binding analysis, he recommends that the free ligand concentration be maintained between 20 and 80% of the total ligand concentration.

Mn^{2+} binding studies of the type described above with purified kidney $(Na^+ + K^+)$-ATPase (21) and Mg^{2+}-ATPase (23) have yielded biphasic Scatchard plots in each case (Figure 1). The binding parameters obtained from computer fitting of the data for both these systems shown in Table I include a single high affinity site per enzyme and a large number of weaker binding sites. The suggestion has been made in both cases (21, 23) that the low affinity sites are metal binding sites on the lipid membrane. In one of these cases (21), this has been confirmed by Mn^{2+} binding studies with delipidated preparations of enzyme.

Table I

Stoichiometry and Dissociation Constants for Mn^{2+} Binding to Membrane ATPases as Determined by Mn^{2+} EPR.

Enzyme	n	K_D
		μM
$(Na^+ + K^+)$-ATPase[a]	1.0 ± 0.2	0.21 ± 0.02
	24 ± 3	185 ± 20
Mg^{2+}-ATPase[b]	$1.0^c \pm 0.2$	2.0 ± 0.2
	34 ± 4	550 ± 60

[a] from Ref. 21.
[b] from Ref. 23.
[c] assumes molecular weight of 469,000.

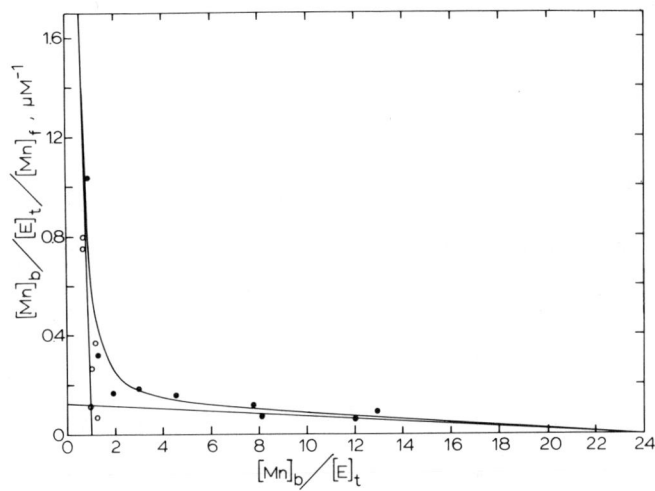

Figure 1A. Scatchard plot of the binding of Mn^{2+} to $(Na^+ + K^+)$-ATPase from sheep kidney medulla (21). In this figure (●) is the native enzyme, (○) is the delipidated enzyme, and (———) assumes that $n_1 = 1$, $K_1 = 0.21\mu M$, $n_2 = 24$, and $K_2 = 185 \mu M$. The free Mn^{2+} was determined by EPR. The solid curves were fitted to the data by computer.

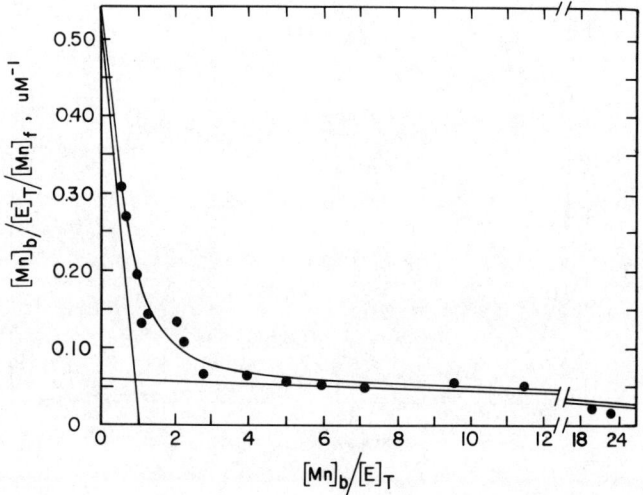

Figure 1B. Scatchard plot of the binding of Mn^{2+} to Mg^{2+}-ATPase from sheep kidney medulla (23). The free Mn^{2+} was determined by EPR. The solid curves were fitted to the data by computer.

B. Active Site Conformations. Electron paramagnetic resonance (EPR) spectra of the bound Mn^{2+} ion can provide insight into molecular motion and structure at the active site of enzyme-Mn^{2+} complexes. Correlation of parameters from the EPR spectra with data obtained from NMR studies of ligands, substrates, and metal-bound water molecules can provide unique information on the catalytic centers of enzymes. However, successful Mn^{2+} EPR studies of Mn^{2+}-enzyme complexes require high concentrations of enzyme, not only for the sake of obtaining reasonably intense signals, but also to suppress the concentration of free Mn^{2+}, so as to minimize the spectral interference from the strong, isotropic signal for the free ion. For this reason, membrane bound enzymes present a potential problem. The particulate suspension state of most membrane enzymes places a severe limit on obtainable enzyme concentrations, while the viscous nature of concentrated suspensions of these enzymes presents problems of sample handling and transfer.

At least in the case of the $(Na^+ + K^+)$-ATPase, however, these problems have been circumvented. As shown in Figure 2, the X-band spectrum of kidney $(Na^+ + K^+)$-ATPase exhibits a "powder" line shape typical of macromolecule complexes with Mn^{2+}. The central portion of the spectrum represents the $-1/2 \longleftrightarrow 1/2$ fine structure transition in which partial resolution of the ^{55}Mn nuclear hyperfine structure is discernible. The spectrum also has a broad component to the low-field side of the main pattern. This low-field signal is part of the fine structure splitting which arises whenever there are asymmetries in the electronic environment of the bound Mn^{2+} (30). Computer simulations suggest that these spectra arise from rhombic distortions of the bound Mn^{2+} ion. On the other hand, the EPR spectrum of the binary Mn^{2+}-ATPase complex at 35 GHz (Figure 2) is much narrower than the corresponding spectrum at 9 GHz. Such a narrowing has been observed routinely for protein-Mn^{2+} complexes (30,31,32). The effects of various nucleotide substrates and substrate analogs on the Mn^{2+} EPR spectrum are depicted in Figures 3, 4 and 5. ATP, ADP, AMP-PNP and (not shown) high concentrations of inorganic phosphate all broaden the hyperfine lines of the X-band Mn^{2+} spectrum, consistent with a change in coordination geometry of the bound Mn^{2+}, a change in accessibility of the Mn^{2+} site to solvent, or both. These spectra also demonstrate that this single, catalytic, enzyme-bound Mn^{2+} ion at the active site does not enter into a Mn^{2+}-ATP complex of the usual type on this enzyme. This may in turn provide evidence that ATP itself can be a suitable substrate for the ATPase.

On the other hand, the effects of the AMP on the Mn^{2+} EPR signal are drastically different from those produced by other phosphate substrates. Instead of the line broadening observed earlier, sharpening of the hyperfine lines is observed in the

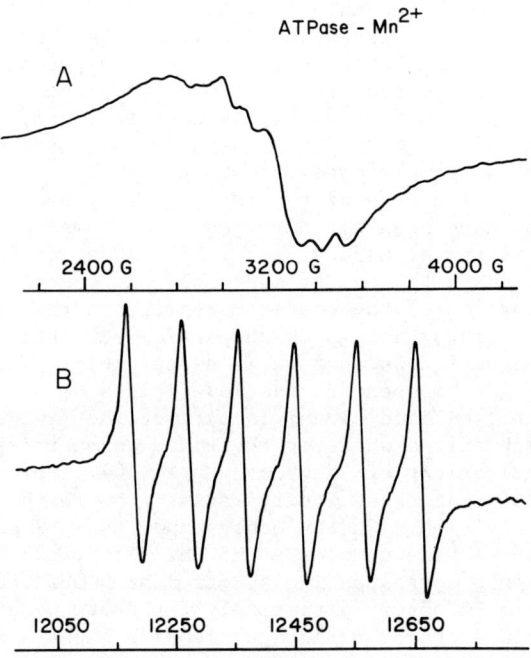

Figure 2. Electron paramagnetic resonance spectra of Mn^{2+} bound to the single catalytic site on $(Na^+ + K^+)$-ATPase. The X-band spectrum (9.5 GHz) is shown in A, while the K-band spectrum (35 GHz) of the same complex is shown in B. The enzyme–Mn^{2+} complex was centrifuged out of 20mM Tes-TMA, pH 7.5, and then combined with buffer so that the final concentrations were: 0.15mM $(Na^+ + K^+)$-ATPase, 0.1mM $MnCl_2$, 20mM Tes-TMA, pH 7.5. T = 23°C.

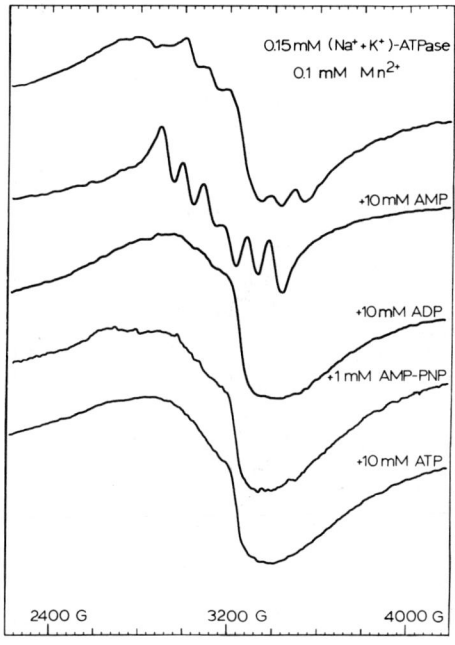

Figure 3. X-band EPR spectra of Mn^{2+} complexes with $(Na^+ + K^+)$-ATPase (21). All solutions contained 20mM Tes-TMA, pH 7.5, 0.15mM ATPase, 0.1mM $MnCl_2$, and the concentrations of the substrates shown.

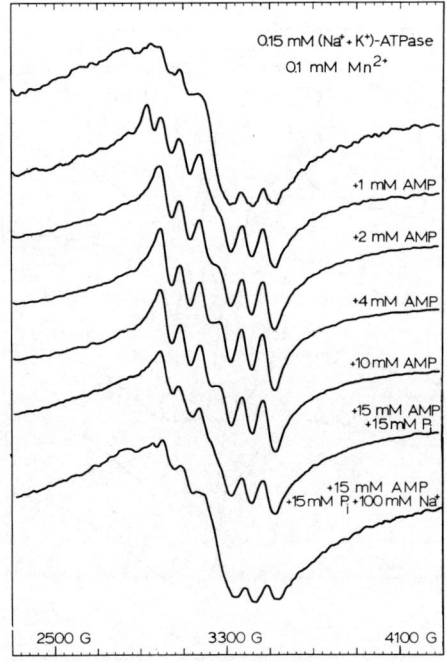

Figure 4. X-band EPR spectra for Mn^{2+} complexes of $(Na^+ + K^+)$-ATPase and AMP (21). Conditions were the same as in Figures 2 and 3, with the concentrations of AMP, inorganic phosphate, and sodium chloride shown.

case of AMP. This can be seen in Figure 4. At 1mM concentration of AMP, seven lines are observed. The six hyperfine lines corresponding to the central $1/2 \longleftrightarrow -1/2$ fine transition narrow significantly, and the low-field line which is part of the fine structure splitting of the Mn^{2+} ion is very prominent. At concentrations of AMP greater than 1 mM this low-field signal disappears, possibly as a result of a greatly reduced axial distortion of the enzyme-metal complex. The hyperfine lines narrow further at a concentration of AMP of 4 mM. Further addition of the substrate to the enzyme system (greater than 4 mM concentration of AMP) produces little or no further effect on the manganese ion spectrum.

The additional effects of inorganic phosphate and Na^+ ion on the enzyme-Mn^{2+}-AMP complex are also shown in Figure 4. Addition of 15 mM inorganic phosphate to the enzyme suspension at the end of the AMP titration causes a partial broadening of the hyperfine lines. However, the addition of sodium ion to this solution results in a further broadening of the hyperfine lines and, more significantly, the reappearance of the broad, low-field line, which is normally observed only in the spectrum of the binary enzyme-Mn^{2+} complex. Whereas phosphate or AMP, acting separately, both cause the disappearance of this line and alterations of the hyperfine lines, AMP plus phosphate plus sodium yield an ATPase-Mn^{2+} spectrum which is indistinguishable from that of the binary complex. The response of the 35 GHz Mn^{2+} EPR spectrum to added ATP is shown in Figure 5. The spectrum is relatively unaffected by ADP and AMP, but ATP (and AMP-PNP) both effect a dramatic broadening of the spectrum. When the system is saturated with either of these latter reactants, the resulting Mn^{2+} spectrum is so broad that it "disappears" and is essentially unobservable at the same conditions of concentration and spectrometer gain. The magnitude of this broadening at 35 GHz is unusual in the literature of Mn^{2+}-protein interactions. The spectra are similar to those observed by Villafranca et al. (32) for complexes of glutamine synthetase-Mn^{2+}-methionine(SR)-sulfoximine-Mg^{2+}-ATP, but even in this latter case, the broadened Mn^{2+} spectrum is still sufficiently narrow to be observed.

The spectra for the ATPase-Mn^{2+} complexes with ATP or AMP-PNP suggest that Mn^{2+} is in a greatly distorting environment. If the substrate-induced distortions were purely axial, a doublet pattern of splitting would be expected for each $-1/2 \longleftrightarrow 1/2$ transition at K-band, whereas more complex effects are anticipated for rhombic distortions. While the spectra for the ATPase-Mn^{2+}-ATP and ATPase-Mn^{2+}-(AMP-PNP) complexes demonstrate that the bound Mn^{2+} is in a rhombic environment, more detailed analyses are not possible.

An estimate of the dissociation constant for ATP at the site involved in broadening of the Mn^{2+} spectrum would facilitate comparisons of structural and mechanistic phenomena in this system. Various studies have provided evidence for both high and low affinity sites for ATP on the $(Na^+ + K^+)$-ATPase (22). In all cases the apparent dissociation constant for the high affinity site is approximately $1 \times 10^{-6}M$, while that of the low affinity site is 500-1000 times higher. (Whether these "two" sites are in fact distinct sites or rather a single site, with different affinities for ATP in two different conformational states, has not been determined conclusively.) Titrations for ATP of the type shown in Figure 4 for AMP indicate that no appreciable broadening occurs until the ATP concentration exceeds $80 \times 10^{-6}M$, essentially ruling out high affinity sites for ATP as the source of the Mn^{2+} spectral broadening.

Our recent studies have also demonstrated the utility of Mn^{2+} EPR studies for measuring metal-substrate distances on $(Na^+ + K^+)$-ATPase. Previous studies have established that paramagnetic CrATP binds to $(Na^+ + K^+)$-ATPase (33), interacts with a K^+ site observed by $^7Li^+$ NMR, and competes with ATP and MgATP at the nucleotide sites on the enzyme (34). If the binding sites for CrATP and Mn(II) are close together on the enzyme, binding of CrATP should effect a diminution of the enzyme-bound Mn(II) EPR spectrum. Such an effect is displayed in the spectra of Figure 6. The addition of β,γ-bidentate CrATP to a solution of Mn(II) and $(Na^+ + K^+)$-ATPase results in a decrease in the amplitude of the spectrum with no apparent change in linewidth. The effect saturates at high levels of CrATP, as shown in a plot of signal intensity versus CrATP in Figure 7. Similar titrations with α,β,γ-tridentate CrATP (Figure 8) had no effect on the Mn(II) signal amplitude, indicating either that tridentate CrATP does not bind to the enzyme or that it binds in a manner which precludes a dipolar interaction with enzyme-bound Mn(II). Titrations with diamagnetic β,γ-bidentate $Co(NH_3)_4ATP$ likewise produced no effect on the Mn(II) EPR spectrum (Figure 8), demonstrating that the effect observed with CrATP is purely a dipolar interaction. The latter experiment is a critical one, in view of the dramatic effect of ATP on the Mn(II) spectrum (Figure 5). Titrations of similar Mn(II) solutions in the absence of enzyme showed no decrease in amplitude of the Mn(II) EPR signal. In the absence of enzyme, the Mn(II) and CrATP species interact only infrequently in solution. On the enzyme, however, these two species are bound rigidly and can interact strongly.

The decrease in the amplitude of the Mn^{2+} EPR signal measured in the presence of CrATP can be used to calculate a Mn^{2+}-Cr^{3+} distance as previously described (35,36). The relevant equation, first described in this form by Leigh (35), is given below; the essence of the distance calculation is the determination of the dipolar interaction constant C. In this equation δH is the linewidth of the

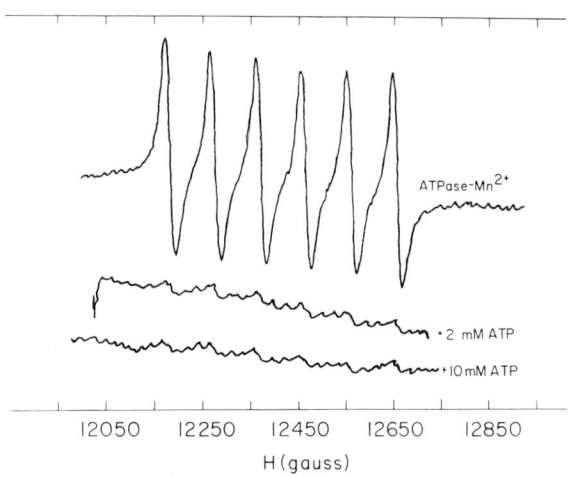

Figure 5. K-band EPR spectra for Mn^{2+} complexes of $(Na^+ + K^+)$-ATPase and ATP. Conditions were the same as in Figures 2 and 3, with the concentrations of ATP shown.

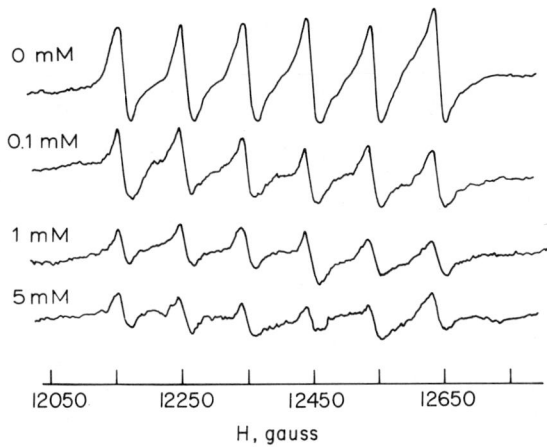

Figure 6. K-band EPR spectra for Mn^{2+} complexes of $(Na^+ + K^+)$-ATPase and β, γ-bidentate CrATP. Conditions were the same as in Figures 2 and 3, except pH = 7.0, with the concentrations of CrATP shown.

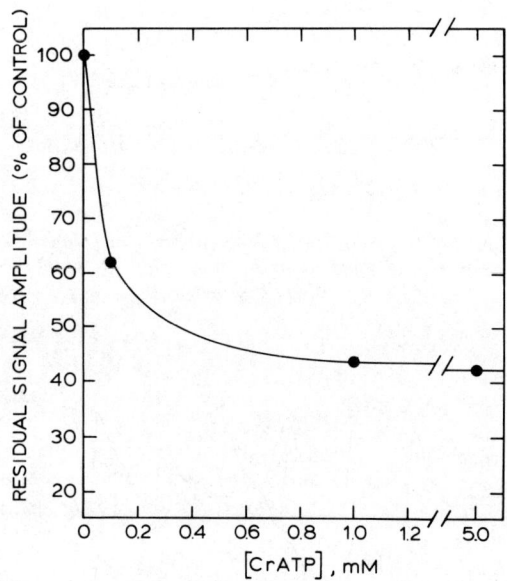

Figure 7. Effect of β, γ-bidentate CrATP on the signal intensity of the Mn^{2+}–$(Na^+ + K^+)$-ATPase complex

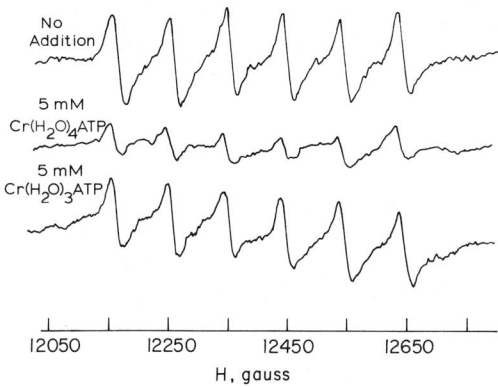

Figure 8A. K-band EPR spectra for Mn^{2+} complexes of $(Na^+ + K^+)$-ATPase and α, β, γ-tridentate CrATP Co(NH_3)$_4$ATP. Conditions were the same as for Figure 6.

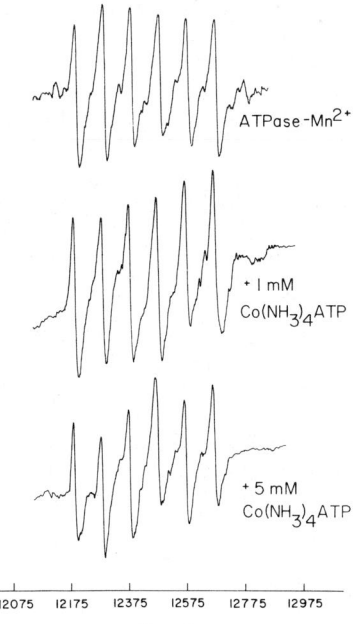

Figure 8B. K-band EPR spectra for Mn^{2+} complexes of $(Na^+ + K^+)$-ATPase and Co(NH_3)$_4$ATP. Conditions were the same as for Figure 6.

$$\delta H = C(1 - 3\cos^2\theta_R')^2 + \delta H_o \qquad (3)$$

observed spin in the absence of dipolar broadening, while θ_r' is the angle between the vector joining the two interacting spins and the applied magnetic field. C is a coefficient describing the interaction and it is defined as:

$$C = g\beta\mu^2\tau_c/hr^6 \qquad (4)$$

In practice it is necessary to perform a computer simulation of the lineshape and signal amplitudes arising for various values of C. A plot of such simulated amplitudes as a function of $C/\delta H_o$ is given in Figure 3 of Leigh (35). For the present case, the data of our Figure 6 yield an interaction constant C of 22.5 gauss. From equation 3, we then calculate a Mn(II)-Cr(III) distance of 8.1 Å, using a correlation time τ_c of 2.7×10^{-10} sec. This value has been determined from the frequency dependence of water proton relaxation in solutions of the ATPase, CrATP and Mg(II) (33).

Gd(III) as a Probe in NMR and EPR Studies of Sarcoplasmic Reticulum Ca^{2+}-ATPase

In order to characterize the active site structure of Ca^{2+}-ATPase from sarcoplasmic reticulum, we have employed Gd^{3+} as a paramagnetic probe of this system in a series of NMR and EPR investigations. Gadolinium and several other lanthanide ions have been used in recent years to characterize Ca^{2+} (and in some cases Mg^{2+}) binding sites on proteins and enzymes using a variety of techniques, including water proton nuclear relaxation rate measurements (35,36,37), fluorescence (38) and electron spin resonance (39). In particular Dwek and Richards (35) as well as Cottam and his coworkers (36,37) have employed a series of nuclear relaxation measurements of both metal-bound water protons and substrate nuclei to characterize the interaction of Gd^{3+} with several enzyme systems.

If gadolinium ion, Gd^{3+}, is to be useful as a spectroscopic probe of Ca^{2+} sites on Ca^{2+}-ATPase, it must first be demonstrated that Gd^{3+} can compete with Ca^{2+} at these Ca^{2+} activator sites. We have examined the effect of Gd^{3+} on the formation of phosphoenzyme from ATP. Both Ca^{2+} and Gd^{3+} stimulate formation of the phosphoenzyme, but different steady-state levels of E-P are formed in these two cases. The effect of Gd^{3+} on Ca^{2+} ion induced E-P formation is shown in Figure 9. Low levels of Gd^{3+} cause a sharp decrease in E-P levels. The data were compared with theoretical curves by assuming a variety of ratios for the dissociation constants of Ca^{2+} and Gd^{3+} from the calcium activator sites. As can be seen in Figure 9, the best fit to the data is

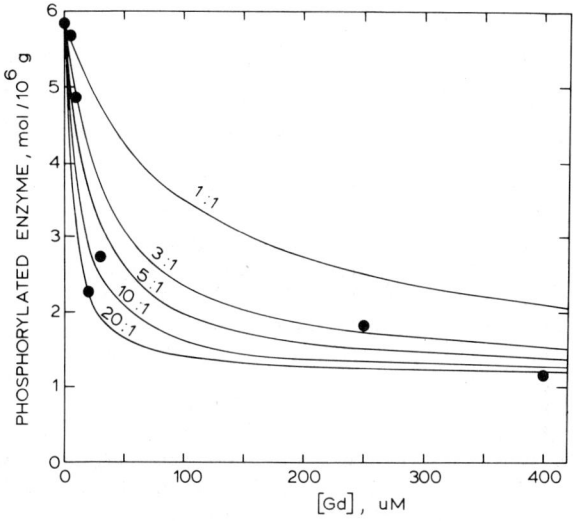

Biochemistry

Figure 9. Steady-state levels of P-E at 0°C as a function of $GdCl_3$ concentration (24). Reaction mixture contained 50mM tris-HCl, pH 7.0, 100mM KCl, 100 μM $CaCl_2$, 10 μM [γ-^{32}P]ATP, and 0.3 mg/mL Ca^{2+}-ATPase. A preincubation for 20 min at 25°C was followed by an incubation of 30 min at 0°C. The reaction was initiated then by addition of [γ-^{32}P]ATP and stopped by addition of 5% trichloroacetic acid. The precipitate was retained and washed on Millipore filters and counted for ^{32}P. The curves shown are calculated by assuming the ratios shown for $K_D Ca^{2+}/K_D^{Gd^{3+}}$.

given by a value for $K_D^{Ca^{2+}}/K_D^{Gd^{3+}}$ of approximately 10. These results indicate that Gd^{3+} binds much more tightly to the Ca^{2+} sites than Ca^{2+} itself and, taken together with the previously determined K_m for Ca^{2+} in the ATPase reaction of 0.35 µM (<u>41</u>), would place the apparent K_D for Gd^{3+} at ~ 3.5 x 10^{-8}M.

A similar, high affinity of Ca^{2+}-ATPase for Gd^{3+} ion was observed in $^7Li^+$ NMR studies. As shown in Figure 10, addition of Gd^{3+} to solutions of LiCl and Ca^{2+}-ATPase produces a dramatic increase in the longitudinal relaxation rate, $1/T_1$, of the $^7Li^+$ nucleus which is proportional to Gd^{3+} concentration until the Gd^{3+} concentration is equal to that of the enzyme. Beyond this point, an additional, linear, but smaller increase in $1/T_1$ of $^7Li^+$ is observed. At ~ 2.25 Gd^{3+} ions/100,000-dalton protein monomer the enzyme is saturated with Gd^{3+}, and no further increases in $1/T_1$ are observed beyond this point. The biphasic nature of this plot is consistent with the sequential binding of two Gd^{3+} ions to the ATPase, with the first of these Gd^{3+} sites exerting a larger paramagnetic effect on enzyme-bound Li^+. Competition studies indicate that Ca^{2+} ion produces a large decrease in $1/T_1$ of $^7Li^+$, consistent with the displacement of Gd^{3+} from the Ca^{2+} sites on the enzyme. Analysis of this data yields a value for $K_D^{Ca^{2+}}/K_D^{Gd^{3+}}$ of 10, in good agreement with the phosphoenzyme data described above.

The binding of Gd^{3+} to Ca^{2+}-ATPase was also examined using water proton nuclear relaxation rates. Figure 11 shows the behavior of the observed enhancement of the longitudinal water proton relaxation rate when Gd^{3+} is used to titrate a solution of the Ca^{2+}-ATPase. At the lower concentrations of Gd^{3+} the large observed enhancement of the water proton relaxation rate suggests the formation of a tight binary Gd^{3+}-ATPase complex. As the concentration of Gd^{3+} is increased, the enhancement decreases toward unity as the Gd^{3+} sites on the enzyme become saturated and the contribution of free Gd^{3+} becomes the predominant term in eq. 5.

$$[Gd]_T \varepsilon^* = [Gd]_F \varepsilon_F + [Gd]_{B1} \varepsilon_{B1} + [Gd]_{B2} \varepsilon_{B2} \tag{5}$$

In this equation the subscripts on the concentration and enhancement terms denote total Gd^{3+} (T), free Gd^{3+} (F), and Gd^{3+} bound at site 1 (B1) or Gd^{3+} bound at site 2 (B2). In this paper we will define site 1 as the site which binds Gd^{3+} more tightly. The data of Figure 11 are consistent with an ε_{B1} of 9.4 and an ε_{B2} of 5.4. When solutions containing 10 or 50 µM Gd^{3+} were titrated with Ca^{2+}-ATPase, the enhancement increased as the enzyme concentration increased. The reciprocal of the observed enhancement was plotted against the reciprocal of the total ATPase concentration (Figure 12) yielding a linear behavior, except at high levels of enzyme where a sharp increase in the observed enhancement is found. This behavior is consistent with two environments for bound Gd^{3+} ion on the Ca^{2+}-

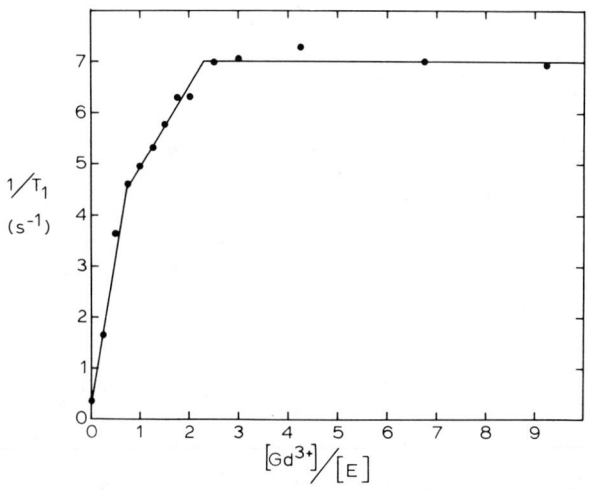

Figure 10. Effect of Gd^{3+} on $1/T_1$ of $^7Li^+$ in the presence of Ca^{2+}-ATPase (24). Solutions contained 0.1M LiCl, 0.05M TMA-Pipes, pH 7.0, and 1.6×10^{-4}M Ca^{2+}-ATPase.

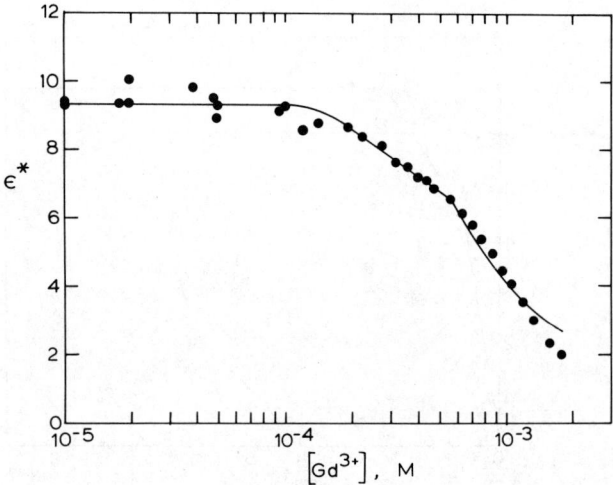

Figure 11. Effect of Gd^{3+} on the enhancement of the longitudinal water proton relaxation rate in solutions containing Ca^{2+}-ATPase (24). The solutions contained 0.05M TMA-Pipes, pH 7.0, 0.24mM Ca^{2+}-ATPase, and the noted concentrations of $GdCl_3$. T = 23°C. The theoretical curve was fitted to the points by assuming that the sites were filled in the manner of the data of Figure 10 and assuming the enhancements described in the text.

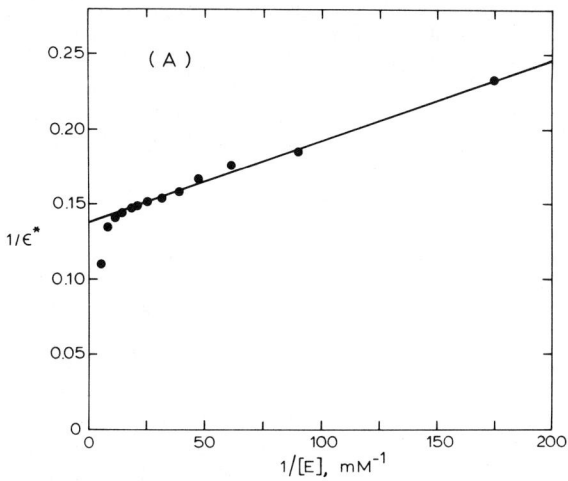

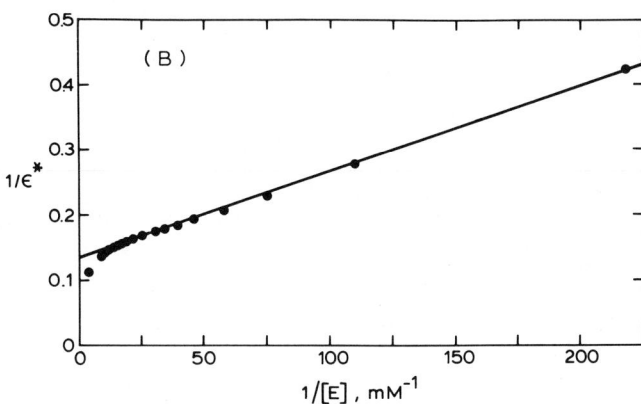

Figure 12. Titration of $GdCl_3$ with Ca^{2+}-ATPase (24). The solutions contained 0.05M TMA-Pipes buffer, pH 7.0, 24.4μM $GdCl_3$ (A) or 49μM $GdCl_3$ (B), and the noted concentrations of Ca^{2+}-ATPase. The value of ϵ^* obtained by extrapolation of the solid line to the infinite protein concentration is the average of the enhancements at Gd^{3+} sites 1 and 2, denoted by ϵ_{B1} and ϵ_{B2}, respectively, in the text.

ATPase. At low concentrations of enzyme in Figure 12, the enzyme binds Gd^{3+} at both types of sites and the observed enhancements in the linear portion of the plot include weighted contributions from the bound enhancements at the two Gd^{3+} sites, as well as from the enhancement of free Gd^{3+} ion in the solution ($\varepsilon_F = 1.0$ by definition). The limiting enhancement of 7.4 at infinite enzyme, which is extrapolated from the linear portion of the plot represents an average of the bound enhancements at the two Gd^{3+} sites. However, as the enzyme concentration is increased the nature of the system changes from enzyme limited to metal limited and the first Gd^{3+} site, which is the tighter binding of the two Gd^{3+} sites and which also displays the larger enhancement, ε_{B1}, begins to pull Gd^{3+} away from site 2 (where the enhancement is lower). The net result is an increase in the observed enhancement and a downward deviation for $1/\varepsilon^*$ at high enzyme in Figure 12.

Paramagnetic effects of metals such as those observed here with Gd^{3+} on the Ca^{2+}-ATPase can be used to determine distances, r, between the metal and pertinent nuclei or, if r is known, to determine q, the number of nuclei involved in the interaction with the metal (11). Thus, the water relaxation data described here can be used to determine the number of exchangeable water protons on Gd^{3+} at the two Ca^{2+} sites on the ATPase. As shown in Table II, the value of q appears to be distinctly different at the two Gd^{3+} sites. Calculations yield three exchangeable water protons on Gd^{3+} at site 1 and two exchangeable protons at site 2.

Table II

Analysis of Frequency Dependence of T_{1p} in Solutions of Ca^{2+}-ATPase and Gd^{3+} ion.

Complex	τ_c^a (s × 10^9)	frequency (MHz)	$1/f(T_{1p})$ (s^{-1}×10^{-6})	$f(\tau_c)$ (s×10^9)	q
ATPase-$(Gd^{3+})_{site\ 1}$	1.91	40	8.45±0.8	4.66	3.11±0.1
		100	4.15±0.4	2.35	3.03±0.1
ATPase-$(Gd^{3+})_{site\ 2}$	2.12	40	4.78±0.4	4.95	1.66±0.1
		100	2.86±0.3	2.29	2.14±0.2

^{a}Determined from the frequency dependence of water proton relaxation (24).

From the frequency dependence of the effect of the enzyme-bound Gd^{3+} on $1/T_1$ of H_2O (24) the correlation times for sites 1 and 2 are found to be 1.9×10^{-9} and 2.1×10^{-9}s, respectively.

These values are too short to be influenced significantly by τ_r, the rotational correlation time of the enzyme-Gd^{3+} complex, or τ_m, the mean residence time of water molecules in the first coordination sphere of the metal. Moreover, the minima in the plots of T_{1p} vs. ω_I^2 indicate that τ_c must be dominated by τ_s, the electron spin relaxation time. The τ_s values for Gd^{3+} in this system are longer than most of those determined previously for Gd^{3+}. The electron spin relaxation time for aqueous Gd^{3+} is (4-7) x 10^{-10}s at 30 MHz (42), while values for τ_s of (2-7) x 10^{-10}s have been reported for complexes of Gd^{3+} with pyruvate kinase (37) and a value of 2.2 x 10^{-10}s has been found for a Gd^{3+}-lysozyme complex (36). Moreover, we have estimated a τ_c of 6.8 x 10^{-10}s for Gd^{3+} bound to parvalbumin.[5] The long Gd^{3+} correlation times found in the present study are consistent with a poor accessibility of these Gd^{3+} sites to solvent water molecules. The electron spin relaxation time, τ_s, is given by eq. 5 (42) where τ_v is a correlation time which is related to the rate at

$$\frac{1}{\tau_s} = B \frac{\tau_v}{1 + \omega_s^2 \tau_v^2} + \frac{4\tau_v}{1 + 4\omega_s^2 \tau_v^2} \qquad (5)$$

which the zero-field splitting is modulated by impact of the solvent molecules on the complex and B is a constant containing the value of the electronic spin and the zero-field splitting parameters. This theory assumes that $\tau_v \ll \tau_s$ and also that the electron spin-lattice relaxation time is much larger than the electron spin-spin relaxation time. The latter assumption will be considered further below. The values of τ_c (i.e., τ_s) obtained for Gd^{3+} bound to the Ca^{2+}-ATPase are 3-10 times longer than those observed with the Gd^{3+} aquo cation and with the protein complexes cited. Since the impact of outer sphere solvent molecules on the Gd^{3+} complex provides the predominant mechanism for electron spin-lattice relaxation, the long values for τ_s found here are thus consistent with a reduced accessibility of solvent water to the Gd^{3+} ion in the ATPase complex.

The possibility of using the electron paramagnetic resonance properties of Gd^{3+} to probe its environment in and interactions with biological molecules has previously received little attention in the literature (40). However, the possibility exists that Gd^{3+} will be a sensitive EPR probe for characterizing macromolecular biological systems such as the Ca^{2+}-ATPase. The EPR spectra of Gd^{3+}, which has S = 7/2. in neutral water and in two different buffers are shown in Figure 13A. The linewidths were found to be independent of pH over the usable range of these buffers and independent of temperature between 4 and 30°C. The spectrum of Gd^{3+} in neutral water is centered around 3248 G, with a linewidth of 530 G. As shown, Gd^{3+} in Pipes buffer, but not in Tes buffer, yielded a spectrum similar to that of the aqueous Gd^{3+} solution. On this basis, all of our Gd^{3+} EPR and NMR studies

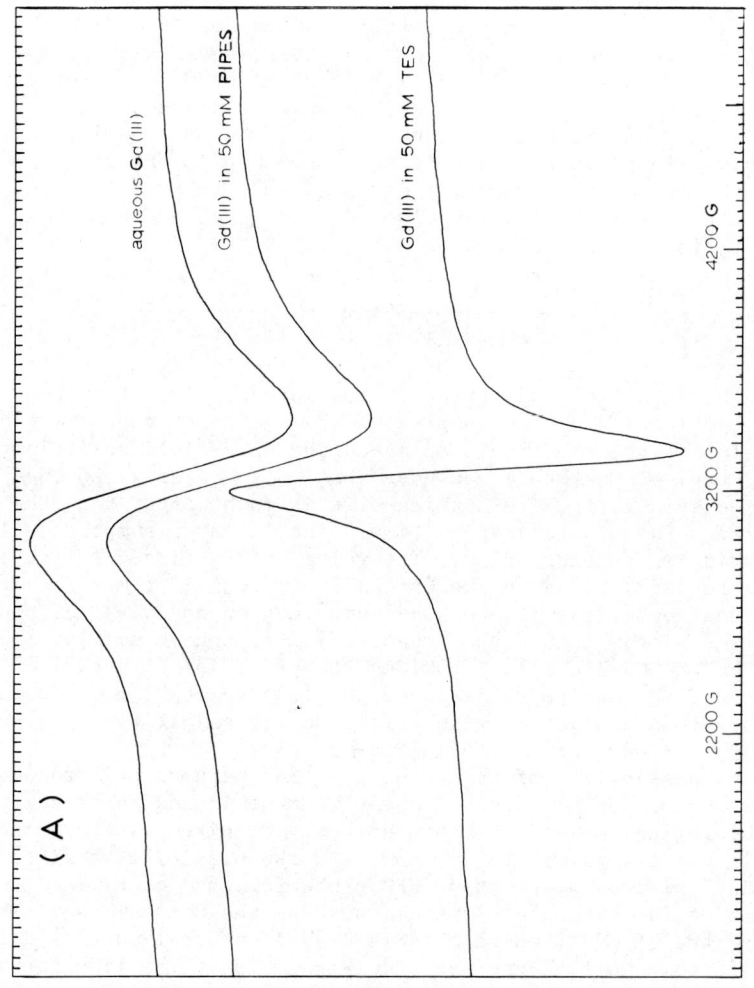

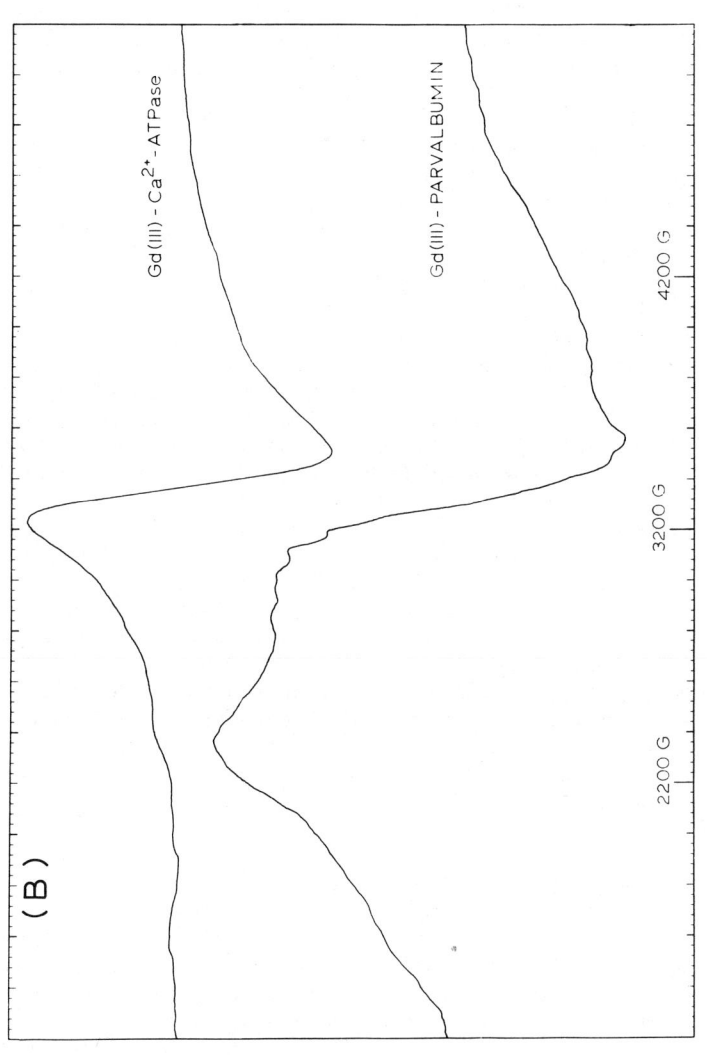

Figure 13. X-band EPR spectrum for Gd^{3+} (A) in aqueous solution in 50mM TMA-Pipes, pH 7.0, and in 50mM TMA-Tes, pH 7.0, with 0.4mM $GdCl_3$ in all cases and (B) bound to Ca^{2+}-ATPase (0.4mM Ca^{2+}-ATPase and 0.38mM $GdCl_3$) or parvalbumin (20mM parvalbumin and 14mM $GdCl_3$) (24). T = 23°C.

were performed in Pipes buffer. As shown in Figure 13B the formation of the Gd^{3+}-ATPase complex results in a shift of the resonance to 3315 G and a decrease in the linewidth to 285 G. Such a decrease in the linewidth may result from a decrease in the Gd^{3+} coordination number upon formation of the macromolecular complex, which could result in greater symmetry and a lower zero-field splitting for the Gd^{3+} ion. This spectrum is independent of temperature between 4 and 25°C and is independent of the Gd^{3+}/ATPase ratio up to 2 Gd^{3+} ions/ATPase molecule. The peak-to-peak linewidth of 285 G sets a lower limit of 2.3×10^{-10}s on the electron spin relaxation time of enzyme-bound Gd^{3+}. This symmetric, narrow EPR spectrum for the Gd^{3+}-ATPase complex is compared in Figure 13B to that of Gd^{3+} bound to parvalbumin, a Ca^{2+}-binding protein from carp. In this case, the spectrum is extremely broad and suggests a greatly distorted Gd^{3+} coordination geometry compared to the Ca^{2+}-ATPase.

Under certain conditions, particularly when the enzyme is incubated with Gd^{3+} at low temperature (2-4°C), the spectrum of the ATPase-Gd^{3+} complex exhibits several low-field transitions, as well as a shoulder to the low-field side of the main transition (Figure 14A). This spectrum is similar to previously reported glassy spectra of transition metal complexes with large zero-field splittings in that they are broad and not centered at g = 2 (56,57,58,59). These spectra are unique in that all have a number of well-defined features at low field (see also ref. (43) and (44) for other powder and glassy spectra of this type). Addition of ATP to the ATPase-Gd^{3+} complex gives rise to additional low field signals (Figure 14B), which are time-dependent. EPR transitions at these low fields normally arise either as the result of dipolar spin-spin interactions (45) or from unusually large zero-field splittings (46). It has been suggested that these time-dependent, low field transitions represent several enzyme states which involve either the close approach of two or more Gd^{3+} ions on the enzyme or a severe change in coordination geometry at the Gd^{3+} sites or both (47).

Summary

It is clear that Mn^{2+} and Gd^{3+} can be sensitive EPR probes of structure and function in membrane enzyme systems. If problems of purity and concentration can be overcome, several other membrane enzymes should be amenable to EPR investigations of the type described here. Moreover, the combination of nuclear relaxation studies and EPR studies of free and bound metals can provide sufficient data for the construction of models of active sites for membrane enzymes. For each paramagnetic probe which can be located unambiguously and uniquely at the active site, one more spatial dimension is added to the picture. Thus the identification of a single Mn^{2+} site on the $(Na^+ + K^+)$-ATPase has permitted the determination of three distances between Mn^{2+} and

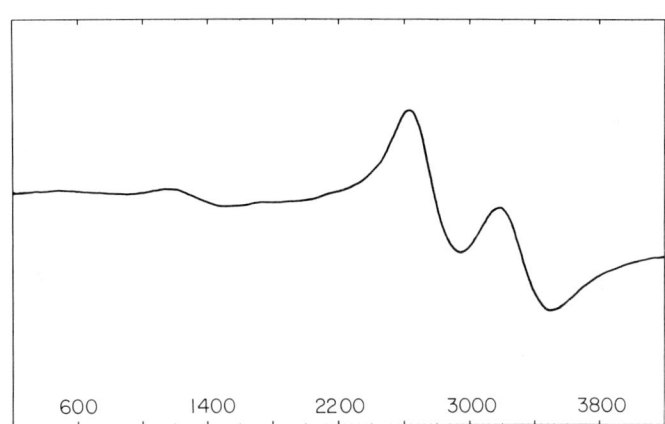

Figure 14A. X-band EPR spectrum of Gd^{3+} complex with Ca^{2+}-ATPase. The solution contained 50mM TMA-Pipes, pH 7.0, 0.32mM Ca^{2+}-ATPase, and 0.32mM $GdCl_3$. The enzyme was maintained at 4°C during the addition of $GdCl_3$ and the running of the spectrum.

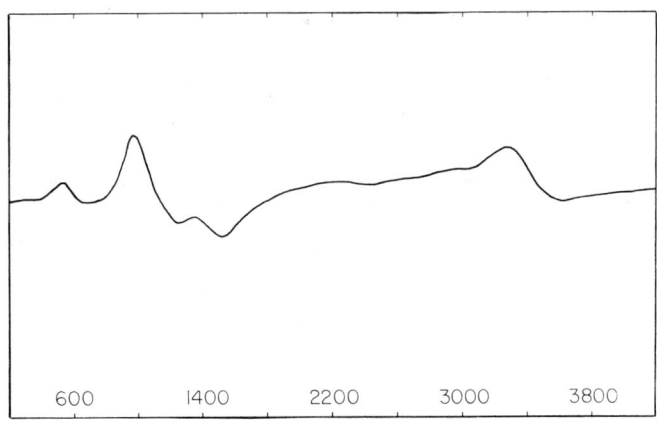

Figure 14B. X-band EPR spectrum of Gd^{3+} complex with Ca^{2+}-ATPase in the presence of ATP. The solution contained 50mM TMA-Pipes, pH 7.0, 0.32mM Ca^{2+}-ATPase, 0.32mM $GdCl_3$, and 0.16mM ATP.

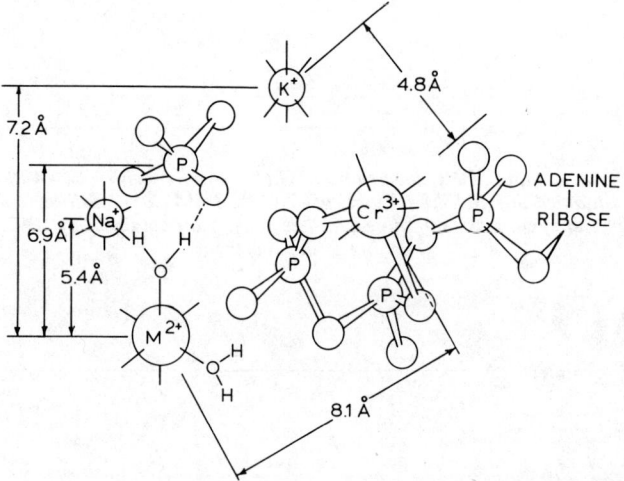

Figure 15. Active-site structure of $(Na^+ + K^+)$-ATPase as determined by ^1H, $^{205}Tl^+$, ^{31}P, and $^7Li^+$ NMR, Mn^{2+} EPR, and kinetic studies

cation or substrate sites on the enzymes (Figure 15). If distances between these sites and the Cr(III) of CrATP can be determined on th enzyme, then it will be possible to construct a two dimensional model for this enzyme. One of these distances, that for Cr(III)-Li$^+$, has already been determined (33) as shown in Figure 15. Experiments in our laboratory have also identified a unique location at the active site of this enzyme for a maleimide spin label (48). Similar distance measurements for this spin label site will provide a third dimension for the model. The assimilation of structural data for the construction of active site models of soluble enzymes has been extensively reviewed (15). For additional discussions of active site models, the reader should consult this and other references on soluble (11,49) and membrane (17-24) enzymes.

Footnotes

The author is a Research Career Development Awardee (NIH-AM00613) of the Public Health Service.

This work was supported by NIH Grants AM19419 and AM00613, ACS-PRF Grant 8757G-4 administered by the American Chemical Society and grants from the Research Corporation, the Muscular Dystrophy Association and the University of Virginia, as well as a grant from the National Science Foundation for the purchase of an EPR spectrometer and computer system.

Abstract

Three membrane-bound adenosine triphosphatase enzymes have been characterized using Mn(II) and Gd(III) electron paramagnetic resonance (EPR) and a variety of NMR techniques. Mn(II) EPR studies of both native and partially delipidated (Na$^+$ + K$^+$)-ATPase from sheep kidney indicate that the enzyme binds Mn^{2+} at a single, catalytic site with $K_D = 0.21 \times 10^{-6}$M. The X-band EPR spectrum of the binary Mn(II)-ATPase complex exhibits a powder line shape consisting of a broad transition with partial resolution of the ^{55}Mn nuclear hyperfine structure, as well as a broad component to the low field side of the spectrum. ATP, ADP, AMP-PNP and P$_i$ all broaden the spectrum, whereas AMP induces a substantial narrowing of the hyperfine lines of the spectrum. The 35 GHz Mn(II) spectrum of this enzyme is severely broadened by ATP and other nucleotide analogs, permitting a characterization of substrate-induced conformational changes at the active site of the (Na$^+$ + K$^+$)-ATPase. In particular we have used the diminution of the Mn(II) EPR spectrum upon titration with β,γ-bidentate CrATP to determine a Mn(II)-Cr(III) distance in the ATPase-Mn-CrATP complex. On the other hand, gadolinium ion, Gd(III), is an effective probe of the Ca(II) sites of the Ca^{2+}-ATPase of sarcoplasmic reticulum. Thus ^{7}Li$^+$ NMR and water proton relaxation studies both indicate binding of two Gd(III) ions to the Ca(II) transport sites of this enzyme. The interactions of these two sites with Gd(III) are not identical. The data indicate three exchangeable water protons on Gd(III) at site 1 with $\varepsilon_b = 9.4$ and two protons at site 2 with $\varepsilon_b = 5.4$. The electron spin relaxation time, τ_s, of Gd(III) at both sites is unusually long (2 ns) and suggests that the Ca^{2+}(II) transport sites are buried within the enzyme-membrane complex. Addition of ATP, AMP-PNP or CrATP results in a decrease in ε^* (consistent with displacement of a water molecule on Gd(III)) as well as EPR spectral changes, which may reflect the events of substrate binding or calcium transport. Mg^{2+}-ATPase, purified from kidney, has been primarily characterized using Mn(II) as a probe in EPR, NMR and kinetic studies. EPR and water proton relaxation rate studies show that the enzyme binds 1 g-ion of Mn(II) per 469,000 g of protein, with a K_D of 2 μM, in good agreement with the kinetically determined activator constant of 3.3 μM. Moreover, the EPR binding studies also indicate the existence of 34 weak sites for Mn(II) per single high affinity Mn(II) site. The K_D for Mn(II) at these sites is 0.55 mM, which agrees well with the low affinity kinetic activator constant for Mn(II) of 0.43 mM, consistent with the additional activation of this enzyme by the large number of weaker Mn(II) binding sites, which appear to be lipid binding sites for divalent metal.

Acknowledgments

I wish to thank Dr. Joseph J. Villafranca for his helpful comments and for the use of the 35 GHz EPR facility at the Pennsylvania State University.

Literature Cited

1. Albers, R. W. Ann. Rev. Biochem., 1967, 36, 727.
2. Rothfield, L., Ed. and Romeo, D. "Structure and Function of Biological Membranes", Academic Press, N.Y., 1971, p. 251.
3. Lin, E.C.C., "Structure and Function of Biological Membranes", L. Rothfield, Ed., Academic Press, N.Y., 1971, p. 286.
4. Rasmussen, H., Goodman, D. and Tenenhouse, A. Critical Rev. Biochem., 1972, 1, 95.
5. Jorgensen, P. L. et Biophys. Acta, 1974, 356, 36.
6. MacLennan, D. J. Biol. Chem., 1970, 245, 4508.
7. Widnell, C. Meth. Enzymol., 1974, 32B, 368.
8. Bock, H. and Fleischer, S. Meth. Enzymol. 1974, 32B, 374.
9. Gantzer, M. L. and Grisham, C. M. Arch. Biochem. Biophys. 1979, 198, 263.
10. Hodges, T. and Leonard, R. Meth. Enzymol. 1974, 32B, 392.
11. Mildvan, A. S. and Cohn, M. Adv. Enzymol. 1970, 33, 1.
12. Reed, G. and Cohn, M. J. Biol. Chem. 1972, 249, 3073.

13. Gupta, R., Fung, C. and Mildvan, A. J. Biol. Chem. 1976, 251, 2421.
14. Balakrishnan, M. and Villafranca, J. Biochemistry 1978, 17, 3531.
15. Mildvan, A. Adv. Enzymol. 1979, 49, 103.
16. Grisham, C. and Mildvan, A. J. Biol. Chem. 1974, 249, 3187.
17. Grisham, C., Gupta, R., Barnett, R. and Mildvan, A. J. Biol. Chem., 1974, 249, 6738.
18. Grisham, C. and Mildvan, A. J. Supramol. Structure, 1975, 3, 301.
19. Grisham, C. M. "Biological Structure and Function", P. Agris, Ed., Academic Press, N.Y., 1978, p. 385.
20. Grisham, C. and Hutton, W. Biochem. Biophys. Res. Comm., 1978, 81, 1406.
21. O'Connor, S. and Grisham, C. Biochemistry, 1979, 18, 2315.
22. Grisham, C. M. "Advances in Inorganic Biochemistry", G. Eichhorn and L, Marzilli, Eds., Elsevier/North-Holland, 1979, p. 193.
23. Gantzer, M. and Grisham, C. Archives of Biochem. and Biophys. 1979, 198, 268.
24. Stephens, E. and Grisham, C. Biochemistry, 1979, 18, 4876.
25. Abragam, A. and Bleaney, B. "Electron Paramagnetic Resonance of Transition Ions", Oxford University Press, London, 1970, p. 491.
26. Hudson, A. and Luckhurst, G. Chem. Rev., 1969a, 69, 191.
27. Hudson, A. and Lewis, J. Trans. Faraday Soc. 1970, 66, 1297.
28. Cohn, M. and Townsend, J. Nature (London) 1954, 173, 1090.
29. Deranleau, D. J. Am. Chem. Soc. 1969, 91, 4044,4050.
30. Reed, G. and Ray, W., Jr. Biochemistry, 1971, 10, 3190.
31. Reed, G. and Cohn, M. J. Biol. Chem. 1970, 245, 662.
32. Villafranca, J., Ash, D. and Wedler, F. Biochemistry 1976, 15, 544.
33. Grisham, C. J. Inorg. Biochem. 1979, manuscript submitted.
34. Gantzer, M., McClaugherty, H., Stephens, E. and Grisham, C. 1979, manuscript in preparation.
35. Leigh, J. J. Chem. Phys. 1970, 52, 2608.
36. Dwek, R. and Richards, R. Europ. J. Biochem. 1971, 21, 204.
37. Cottam, G. Valentine, K., Thompson, B. and Sherry A. Biochemistry, 1974, 13, 3532.
38. Valentine, K. and Cottam, G. Arch. Biochem. Biophys. 1973, 158, 346.
39. Epstein, M., Levitski, A. and Reuben, J. Biochemistry 1974 13, 1777.
40. Geraldes, E. and Williams, R. J. Chem. Soc., Dalton Trans. 1977, 18, 1721.
41. Yamada, S. and Tonomura, Y. J. Biochem. (Tokyo) 1972, 74, 417.
42. Bloembergen, N. and Morgan, L. J. Chem. Phys. 1961, 34, 842.
43. Aasa, R., Falk, K. and Reyes, S. Ark. Kemi 1967, 25, 309.

44. Aisen, P., Aasa, R. and Redfield, J. Biol. Chem., 1969, 244, 4628.
45. Birgeneau, R., Hutchings, M. and Wolf, W. Phys. Rev. 1969, 197, 275.
46. McGeehin, P. and Henderson, B J. Physics C, 1974, 7, 3988.
47. Stephens, E. and Grisham, C. Fed. Proc. 1978, 37, 1483.
48. O'Connor, S. and Grisham C, manuscript in preparation, 1979.
49. Mildvan, A., Sloan, D., Fung, C., Gupta, R. and Melamud, E. J. Biol. Chem. 1976, 251, 2431.

RECEIVED March 4, 1980.

Molecular Structure of Vinyl Chloride–Vinylidene Chloride Copolymers by Carbon-13 NMR

CHARLES J. CARMAN[1]

The B. F. Goodrich Company, Research & Development Center, 9921 Brecksville Road, Brecksville, OH 44141

High resolution ^{13}C nmr is established as perhaps the most discriminating spectroscopic method in recent years for determining the molecular structure of macromolecules. For chlorinated polymers the direct and long range substituent effects of the halogen produce unique ^{13}C chemical shifts that enable the determination of tacticity in PVC[1], average block lengths in vinyl chloride-butadiene copolymers[2], and monomer composition and sequence distribution in a variety of vinyl chloride copolymers[3]. In addition, Keller et al have attempted to apply interpretational analogies to more complicated chlorinated polyalkanes[4]. The present ^{13}C nmr analysis of copolymers of vinyl chloride (VCl) and vinylidene chloride (VCl$_2$) demonstrates the value of using internal consistencies for determining monomer composition and sequence distribution.

Results and Discussion

Figure 1 shows the spectra of three copolymers and compares their spectra to that of a PVC sample. The resonances centered around 46 ppm in Figure 1a are from the CH$_2$ carbon and those resonances in the 54 ppm region are from the CHCl carbon. The fine structure in both the methine and methylene regions is from tacticity information.

At very low levels of vinylidene chloride, the VCl$_2$ monomer will be flanked by VCl monomer. The isolated structure is shown in Figure 2a. The carbon bearing both chlorine atoms has a chemical shift of 87.50 ppm and is identified in Figure 1b. The chemical shift at 52.33 ppm in Figure 1b is assigned to the methylene carbon adjacent to a vinylidene (or dichloro carbon) structure.

---

[1] Current address: B. F. Goodrich Company, Engineered Products Group, 500 S. Main Street, Akron, OH 44318.

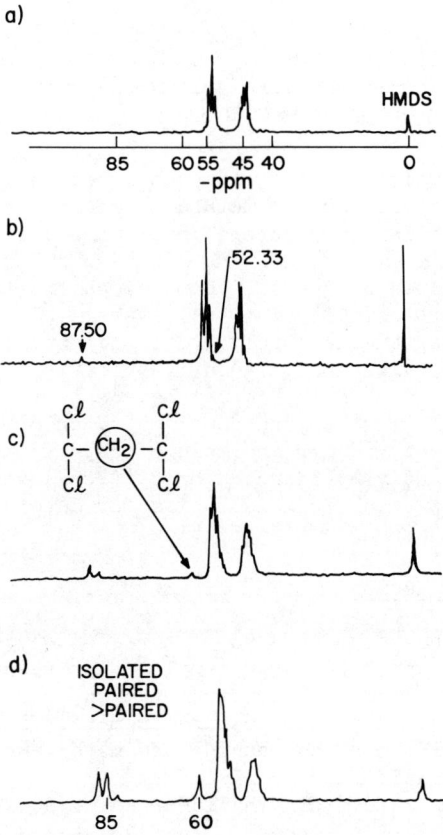

Figure 1. $^{13}C\{^1H\}$ NMR spectra of PVC (a) and VCl–VCl$_2$ Copolymers B (b), C (c), and D (d)

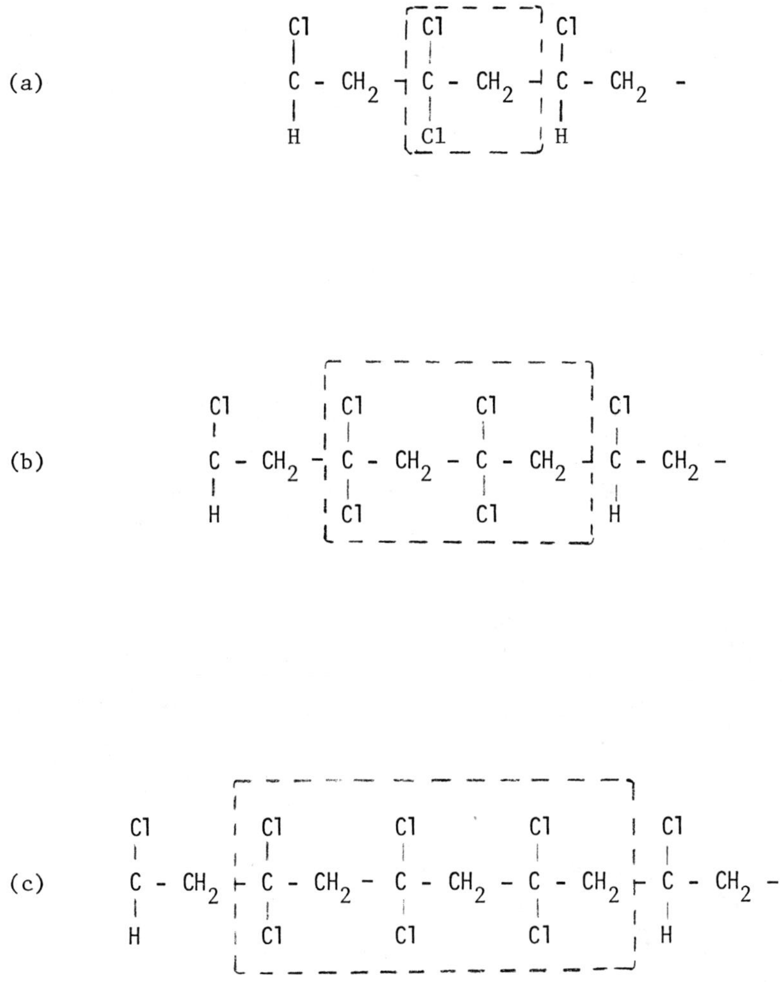

(a)

```
        Cl        ┌ Cl     ┐  Cl
        |         | |      |  |
        C - CH₂ ┐ | C - CH₂|┤ C - CH₂  -
        |       | | |      |  |
        H       | |_Cl_ _ _|  H
```

(b)

```
        Cl        ┌ Cl       Cl         ┐ Cl
        |         | |        |          | |
        C - CH₂ ┐ | C - CH₂ - C - CH₂  ┤ C - CH₂ -
        |       | | |        |          | |
        H       | | Cl       Cl         | H
                  └ _ _ _ _ _ _ _ _ _ _ ┘
```

(c)

```
        Cl      ┌ Cl       Cl        Cl        ┐ Cl
        |       | |        |         |         | |
        C - CH₂ ├ C - CH₂ - C - CH₂ - C - CH₂ ┤ C - CH₂ -
        |       | |        |         |         | |
        H       | Cl       Cl        Cl        | H
                └ _ _ _ _ _ _ _ _ _ _ _ _ _ _ _┘
```

Figure 2. Carbon environment in VCl–VCl₂ copolymers representing (a) isolated, (b) paired, and (c) greater-than-paired structures

The next higher structure formed would be paired VCl_2 monomer. This is shown in Figure 2b. In the paired structure, there are still two methylene carbons which form the VCl_2 and VCl junction. There is also a unique methylene between two CCl_2 carbons at 60.49 ppm and the chemical shift of the CCl_2 in a paired structure at 85.49 ppm. As vinylidene chloride increases in concentration the polymer would form greater than paired sequences; shown in Figure 2c. A unique carbon resonance formed from this structure is the CCl_2 flanked by CCl_2 on either side in the β position. The corresponding chemical shift is shown at 82.83 ppm in Figure 1d. The other carbons in the (> paired) sequence only change the relative areas of the other assignments in the spectrum. A summary of the ^{13}C chemical shift assignments for VCl-VCl_2 copolymers is given in Table I.

To calculate monomer composition, these chemical shift assignments must follow peak area relationships deduced from isolated, paired, and > paired structures.

<u>Isolated</u>: The moles of isolated VCl_2 monomer is proportional to the isolated CCl_2 carbon area. There is also produced an area twice the isolated CCl_2 area which is assigned to the two flanking CH_2 carbons. The PVC CH_2 carbon is reduced in area by an amount equal to the isolated CCl_2 carbon area.

<u>Paired</u>: The moles of paired VCl_2 monomer produce a directly proportionate area from the paired CCl_2 carbon and an equal area from the outer CH_2. There will also be an area equal to half the paired CCl_2 area assigned to the inner CH_2 carbon. The PVC CH_2 area is reduced in area by an amount one half the paired CCl_2 (because half the outer CH_2 area comes from the PVC monomer).

<u>>Paired</u>: The area relationships for greater than paired depend on the lengths of the contiguous units. For a run three units long there is an area from the > paired CCl_2 carbon directly proportional to one third the number of monomers in a contiguous sequence. There are contributions of twice this area to both the inner CH_2 resonance areas. The PVC CH_2 is reduced in area by half the outer CH_2 area. Therefore, for VCl_2 sequences (n) assigned to the > paired CCl_2, the area (n-1) is assigned to the inner CH_2.

Using these area relationships the equations in Table II were written to use the measured peak areas from a copolymer of vinyl chloride to calculate monomer compositions. The equations also provide checks on internal consistency of the spectra. These relationships must hold as the monomer ratio is varied or errors in assignment have been made. Equations (1) and (2) are simply two ways of expressing the CHCl spectral region in terms of contribution of PVC methine. Equation (4) is a means of calculating the concentration of vinyl chloride in terms of its methylene area after correcting for those methylene carbons that have been shifted (ψ, equation 3) under the CHCl resonance region by virtue of being adjacent to a CCl_2 carbon. Equations (5) and (6) are two area relationships that provide the vinylidene chloride concentration.

TABLE I

^{13}C Chemical Shifts[a] for VCl-VCl$_2$ Copolymers

Structure	δC (ppm)
*CHCl[b]	
rr	55.21, 55.12
mr	54.35, 54.19
mmmm	53.48
mmmr	53.35
rmmr	53.22
CHCl*CH$_2$CHCl[b]	
rrr	45.80
rmr	45.41
rrm	45.07
mmr + mrm	44.39
mmm	43.65
*CCl$_2$	
isolated	87.50
paired	85.09
> paired	82.83
CCl$_2$*CH$_2$-R	
R = CHCl	52.33
R = CCl	60.49
> paired	~53-55

[a] relative to hexamethyldisiloxane (HMDS), measured in 1,2,4,-trichlorobenzene at 373° K

[b] reference 1

TABLE II

Equations to Calculate Monomer Composition of
Vinyl Chloride-Vinylidene Chloride Copolymers from ^{13}C NMR Areas

Mole VCl = PVC(CHCl) = (area CHCl)$_{\text{total region}}$ −
[(isolated CCl$_2$)$^{\text{area}}_{\text{VCl}}$ + (isolated CCl$_2$)$^{\text{area}}_{\text{VCl}_2}$ + 0.5 (outer CCl$_2$)$^{\text{area}}_{\text{VCl}}$
+ 0.5 (outer CCl$_2$)$^{\text{area}}_{\text{VCl}_2}$]

(1): mols (VCl) = PVC (CHCl) = (area CHCl)$_{\text{total region}}$ −
2(isolated CCl$_2$ area) − (outer CCl$_2$ area)

(2): mols (VCl) = PVC (CHCl) = (area CHCl)$_{\text{total region}}$ −
[isolated CCl$_2$)$^{\text{area}}_{\text{VCl}}$ + 0.5 (outer CCl$_2$)$^{\text{area}}_{\text{VCl}}$ + (total CCl$_2$)$^{\text{area}}_{\text{VCl}}$ −
(inner CH$_2$)$^{\text{area}}_{\text{VCl}_2}$]

(3): ψ = (total CCl$_2$− inner CH$_2$)^{area} ≡ (isolated CCl$_2$+ 0.5 outer CCl$_2$)^{area}

(4): mols VCl = PVC (CH$_2$) = (area CH$_2$) + ψ
(e.g. shifted VCl (CH$_2$))

(5): mols VCl$_2$ = VCl$_2$ (CCl$_2$) =(area CCl$_2$)$_{\text{total region}}$

| 1 ≡ 2 ≡ 4 or take average |

(6): mols VCl$_2$ = VCl$_2$ (CH$_2$) = (inner CH$_2$)$^{\text{area}}_{\text{VCl}_2}$ + (isolated CCl$_2$)
$^{\text{area}}_{\text{VCl}_2}$ + 0.5 (outer CCl$_2$)$^{\text{area}}_{\text{VCl}_2}$

mol. wt. VCl = 62.5; wt. % Cl = 56.6
mol. wt. VCl$_2$ = 97.0; wt. % Cl = 73.2

| 5 ≡ 6 or take average |

TABLE III
^{13}C NMR Analysis of VCl-VCl$_2$ Copolymers

Sample	Mol % VCl$_2$	Wt. % Cl (^{13}C)	Wt. % Cl (wet)	Relative Triad Distribution[a]			$\overline{n}_0$[b]
				(101)	(001)	(000)	Observed
A	1.7	56.9	55.78, 55.85	1.0	0	0	1.00
B	2.2	57.1	—	1.0	0	0	1.00
C	10.1	59.0	58.67, 58.29	0.63	0.33	0.04	1.26
D	26.6	62.5	—	0.45	0.42	0.12	1.50

[a] VCl is 1; VCl$_2$ is 0

[b] $\overline{n}_0 = \dfrac{N_{101} + N_{001} + N_{000}}{N_{101} + (0.5) N_{001}}$

Four VC1-VCl$_2$ copolymers were analyzed and the results are summarized in Table III. The internal consistency checks from Table II held. The resulting chlorine analysis for sample C agreed very well with the wet analytical results. The discrepancy in sample A suggests the wet results are in error as the reported values were less than the PVC theoretical value.

Using the observed triad concentrations in Table III, the number average sequence lengths for the VCl$_2$ monomer were calculated[5] using the equation:

$$\bar{n}_0 = \frac{N_{101} + N_{001} + N_{000}}{N_{101} + 0.5\, N_{001}}. \tag{1}$$

The observed number average sequence lengths deviate from those predicted from Bernoullian statistics, i.e., $\bar{n}_0 = 1/[\text{VCl}]$. However, corresponding calculations based on first order Markovian statistics are in excellent agreement.

To calculate the number average sequence length using first order Markovian statistics, it is necessary to estimate p_{01}, (for $\bar{n}_0 = 1/p_{01}$). The information that can be measured from the spectra in Figure 1 is monad and triad concentrations: (0), (1), (101), (001), and (000). In this notation vinyl chloride is represented by (1) or vinylidene chloride by (0). Using these measurable structures an estimate for p_{01} was derived using an approach suggested by Randall[6] for the determination of sequence distribution for hydrogenated polybutadienes. This can be accomplished in two ways; from the ratio of (101)/(001) and from the ratio of (101)/(0).

Equations (2) through (5) describe the observable structures in terms of transition probabilities:[5]

$$000 = p_{10}(1-p_{01})^2/(p_{01} + p_{10}) \tag{2}$$

$$101 = (p_{01})^2 p_{10}/(p_{01} + p_{10}) \tag{3}$$

$$001 = 2(p_{01})(p_{10})(1-p_{01})/(p_{01} + p_{10}) \tag{4}$$

$$0 = p_{10}/(p_{01} + p_{10}/(p_{01} + p_{10}) \tag{5}$$

Combing these equations enables one to estimate values for p_{01} from either of two ratios shown as follows:

$$(101)/(001) = p_{01}/2(1-p_{01}) \tag{6}$$

$$(101)/(0) - (p_{01})^2 \tag{7}$$

Table IV shows these ratios for the four VC1-VCl$_2$ copolymers. The value for p_{01} can be calculated using both equations for samples C and D, and are in good agreement. As seen in

TABLE IV

Number Average Sequence Lengths Calculated from ^{13}C NMR Spectra

Sample	$\langle 101\rangle/\langle 001\rangle$	$\langle 101\rangle/\langle 0\rangle$ [a]	1st Order Markovian			Bernoullian [b]	Observed
			p_{01} [c]	p_{01} [d]	$\bar{n}_0$ [e]		
A	—	1.00	—	1.00	1.00	1.02	1.00
B	—	1.00	—	1.00	1.00	1.02	1.00
C	1.909	0.630	0.79	0.79	1.26	1.11	1.26
D	1.071	0.451	0.68	0.67	1.48	1.36	1.50

[a] absolute concentration need for this ratio is product of ⟨101⟩ from Table III and mol fraction VCl$_2$

[b] $\bar{n}_0 = 1/p_1$

[c] calculated using equation 6

[d] calculated using equation 7

[e] $\bar{n}_0 = 1/p_{01}$

Table IV, the resulting calculation of VCl_2 number average sequence length calculated using the first order Markov agrees very well with the observed sequence lengths.

There is another way to confirm that the triad distribution cannot result from a random polymerization. The observed relative proportion of the triad structure in Figure 1d is (1):(.93):(.26). The corresponding distribution for the triad structure for .266 mole fraction VCl_2 would be (1):(.36):(.13). Obviously, this comparison confirms from just a consideration of the triad distribution that the polymer does not conform to Bernoullian statistics.

The data shown in Tables III and IV show that the ^{13}C nmr spectra of vinyl chloride-vinylidene chloride copolymers have a redundancy of structural relationships. By analyzing a range of compositions, this system has been found to yield a reasonable description of both monomer composition and monomer sequence distribution. The data also show that this copolymer is a good example of a system best described by first order Markovian statistics as compared to Bernoullian statistics.

Experimental

The ^{13}C nmr spectra were obtained at 22.6 MHz using a Bruker HX90E spectrometer from 40 percent polymer solutions in 1,2,4-trichlorobenzene at 373K. The spectral conditions were: $\pi/2$ (25μs), 6 kHz sweepwidth, 16 k fid, 1.5 Hz line broadening, 5 sec rep. rate, number of scans to give good S/N, (usually about 6000). Perdeuterobenzene was added as internal lock. Compositions were calculated from electronic integrals.

Acknowledgements

I would like to thank M. L. Dannis for supplying the polymers and R. E. Scourfield for obtaining the nmr spectra. Thanks is also given to The BFGoodrich Company for permission to publish this work.

Abstract

Carbon-13 nuclear magnetic resonance was used to determine the molecular structure of four copolymers of vinyl chloride and vinylidene chloride. The spectra were used to determine both monomer composition and sequence distribution. Good agreement was found between the chlorine analysis determined from wet analysis and the chlorine analysis determined by the ^{13}C nmr method. The number average sequence length for vinylidene chloride measured from the spectra fit first order Markovian statistics rather than Bernoullian. The chemical shifts in these copolymers as well as their changes in areas as a function of monomer composition enable these copolymers to serve as model compounds for making structural assignments in other chlorinated polymers.

Literature Cited

1. C. J. Carman, Macromolecules, 6, 725 (1973).

2. C. E. Wilkes, J. Polym. Sci., Polymer Symposium 60, 161 (1977).

3. F. Keller, Plaste U. Kautschuk, 23, 730 (1976); B. Hosselbarth and F. Keller, Faserforsch. u. Textiltechnik/Z. Polymerforsch., 28, 325 (1971); F. Keller and C. Mugge, ibid, 27, 347 (1976); F. Keller, S. Zepnik, and B. Hosselbarth, ibid, 29, 29 (1978).

4. F. Keller and B. Hosselbarth, Faserforsch. u. Textiltechnik Z. Polymerforsch., 26, 329 (1975); ibid, 27, 453 (1976); ibid, 28 287 (1977); R. Lukas, M. Kolinsky, D. Doskocilova, J. Polym. Sci., 16, 889 (1978).

5. J. C. Randall, "Polymer Sequence Determination: Carbon-13 NMR Method", Academic Press, New York, 1977.

6. J. C. Randall, J. Polym. Sci., Polym. Phys. Ed. 13, 1975 (1975).

RECEIVED May 21, 1980.

Characterization of Long-Chain Branching in Polyethylenes Using High-Field Carbon-13 NMR

J. C. RANDALL

Phillips Petroleum Company, Bartlesville, OK 74004

The simplest polymer molecule examined for both its dynamic and structural characteristics utilizing carbon-13 nuclear magnetic resonance has been polyethylene. At first sight, a polymer molecule which is essentially a "polymethylene" would not seem to possess enough different structural characteristics to warrant more than a casual investigation. However, there are a variety of different polyethylenes available commercially that have substantially different physical properties and, subsequently, different end use applications. The fact that catalysts are sought to produce polyethylenes with desirable properties and that post treatments may be used to alter certain physical characteristics imply that there is a direct link between physical properties and molecular structure. This link between structure and properties and the commercial importance of polyethylene have brought about a need for a thorough structural characterization.

Polyethylenes produced commercially via high pressure, free radical processes have densities around 0.92 g/cc and are referred to simply as "low density" polyethylenes. It has been well established from infrared measurements that these low density polyethylenes possess appreciable quantities of ethyl and butyl branches (1-3) but it was not until C-13 NMR became available that an absolute identification, both qualitatively and quantitatively, of the short branches became possible (4-8). Long chain branching is also present in high pressure process low density polyethylenes and carbon-13 NMR was useful here also in establishing the identity and relative amounts of long versus short chain branches (9-11).

Polyethylenes with densities around 0.96 g/cc are categorized as high density polyethylenes and are prepared using either titanium or chromium based catalysts. These polyethylenes are usually linear although the physical and rheological properties of some high density polyethylenes have suggested the presence of long chain branching (12) at a level one to two orders of magnitude below that found for low density polyethylenes prepared by a high pressure process. A measurement of long chain branching in high density polyethylenes has been elusive because of the concentrations involved (13) and can only be directly provided by recently available, high field, high sensitivity NMR spectrometers. The purpose of this chapter will be to review briefly the history of structural studies of polyethylene and show where these recent advances in C-13 NMR instrumentation have greatly enhanced our knowledge about polyethylene structure.

Among the most important polyethylene structural characteristics are the weight and number average molecular weights, M_w and M_n, and the molecular weight distribution characterized by M_w/M_n. Since C-13 NMR can also be used to measure a number average molecular weight, it may be advantageous to examine some of the molecular weight characteristics of polyethylenes. As shown by the size exclusion (or gel permeation) chromatographs in Figure 1, the molecular weight distributions are generally broad but can be significantly characteristic to distinguish among certain types of polyethylenes. For example, the first polymer in Figure 1 has a relatively narrow molecular weight distribution with an M_w/M_n of approximately three. The second polyethylene has a much broader molecular weight distribution with an M_w/M_n of approximately twenty. Finally, the third polyethylene has a distinctly bimodal molecular weight distribution. These size exclusion chromatograms do serve as distinguishing fingerprints for the molecular weight distributions which can only be measured quantitatively by the ratio, M_w/M_n. Recent improvements in size exclusion chromatography techniques have permitted faster and more reliable molecular weight determinations (14).

Polyethylenes prepared with a Ziegler type, titanium based catalyst have predominantly n-alkyl or saturated end groups. Those prepared with chromium based catalysts have a propensity toward more olefinic end groups. As will be seen later, the ratio of olefinic to saturated end groups for polyethylenes prepared with chromium based catalysts is approximately unity. The end group distribution is, therefore, another structural feature of interest in polyethylenes because it can be related to the catalyst employed and possibly the extent of long chain branching. Infrared has been a useful technique for measuring the various types of olefinic end groups (15), which may be

$-CH_2-CH=CH_2$ "vinyl"

$-CH_2-\underset{CH_3}{C}=CH_2$ "vinylidene"

$-CH_2-CH=CH-CH_3$ "internal cis or trans"

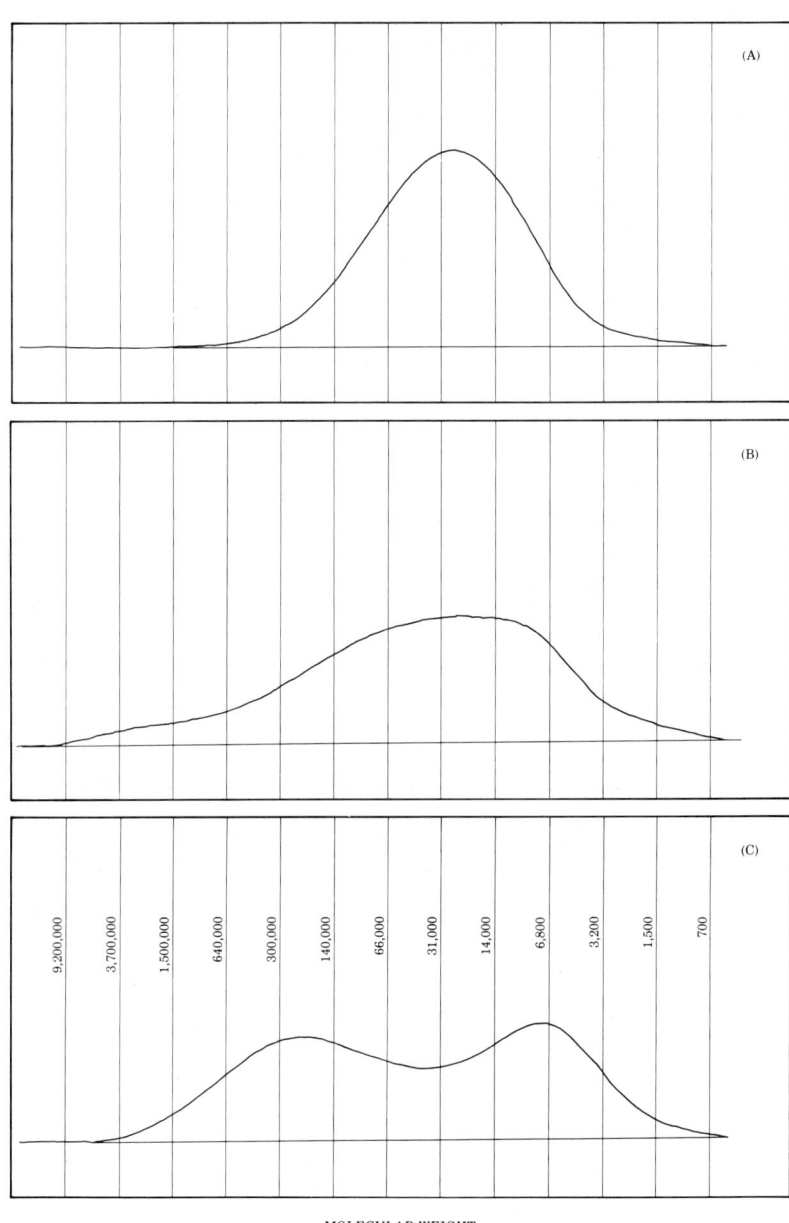

Figure 1. SECs of (A) NBS 1475, (B) Phillips PE 5003, and (C) Hizex 7000

Prior to the availability of C-13 NMR, there was no technique for measuring directly the saturated end group concentration. Now it is possible not only to measure concentrations of saturated end groups, but also the olefinic end groups and, subsequently, an end group distribution.

Short chain branches can be introduced deliberately in a controlled manner into polyethylenes by copolymerizing ethylene with a 1-olefin. The introduction of 1-olefins allows the density to be controlled and butene-1 and hexene-1 are commonly used for this purpose. Once again, as in the case of high pressure process low density polyethylenes, C-13 NMR can be used to measure ethyl and butyl branch concentrations independently of the saturated end groups. This result gives C-13 NMR a distinct advantage over corresponding infrared measurements because the latter technique can only detect methyl groups irrespective of whether the methyl group belongs to a butyl branch or a chain end (16). As will be seen shortly, C-13 NMR also has a disadvantage in branching measurements because only branches five carbons in length and shorter can be discriminated independently of longer chain branches (5)(9). Branches six carbons in length and longer give rise to the same C-13 NMR spectral pattern independently of the chain length. This lack of discrimination among the longer side-chain branches is not a deterring factor, however, in the usefulness of C-13 NMR in a determination of long chain branching.

By far the most difficult structural measurement and, as stated previously, the most elusive, has been long chain branching. Long chain branching in high density polyethylenes has long been considered a factor affecting certain observed physical properties, for example, environmental stress cracking, rheological properties and processing behavior although conclusive proof has been difficult to obtain. In low density polyethylenes, the concentration of long chain branches is such (>0.5 per 1,000 carbons) that characterization through size exclusion chromatography in conjunction with either low angle laser light scattering or intrinsic viscosity measurements becomes feasible (9-11)(13)(17-18). When carbon-13 NMR measurments have been compared to results from polymer solution property measurements, good agreement has been obtained between long chain branching from solution properties with the concentration of branches six carbons long and longer (9)(10). Unfortunately, these techniques utilizing solution properties do not possess sufficient sensitivity to detect long chain branching in a range of one in ten thousand carbons, the level suspected in high density polyethylenes. The availability of superconducting magnet systems has made measurements of long chain branching by C-13 NMR a reality because of a greatly improved sensitivity. An enhancement by factors between 20 to 30 over conventional NMR spectrometers has been achieved through a combination of higher field strengths, 20 mm probes, and the ability to examine polymer samples in essentially a melt state. The data discussed in this study have been obtained from both conventional iron magnet spectrometers with field strengths of 23.5 kG and superconducting magnet systems operating at 47 kG. Let us now begin our discussion of polyethylene long chain branching with an examination of the C-13 NMR structural sensitivity and later turn to quantitative sensitivity and detection limit.

The C-13 NMR spectra from a homologous series of six linear ethylene 1-olefin copolymers beginning with 1-propene and ending with 1-octene are reproduced in Figures 2 and 3. The side-chain branches are, therefore, linear and progress from one to six carbons in length. Also, the respective 1-olefin concentrations are less than 3%; thus, only isolated branches are produced. Unique spectral fingerprints are observed for each branch length. The chemical shifts, which can be predicted with the Grant and Paul parameters (5)(19) are given in Table I for this series of model ethylene-1-olefin copolymers. The nomenclature, used to designate those polymer backbone and side-chain carbons discriminated by C-13 NMR, is as follows:

$$\begin{array}{c} \gamma\beta\alpha\alpha\beta\gamma \\ -CH_2-CH_2-CH_2-CH_2-CH-CH_2-CH_2-CH_2-CH_2-CH_2- \\ | \\ 5\ CH_2 \\ | \\ 4\ CH_2 \\ | \\ 3\ CH_2 \\ | \\ 2\ CH_2 \\ | \\ 1\ CH_3 \end{array}$$

The distinguishable backbone carbons are designated by Greek symbols while the side-chain carbons are numbered consecutively starting with the methyl group and ending with the methylene carbon bonded to the polymer backbone (5). The identity of each resonance is indicated in Figures 2 and 3. It should be noticed in Figure 3 that the "6" carbon resonance for the hexyl branch is the same as α, the "5" carbon resonance is the same as β, and the "4" carbon resonance is the same as γ. Resonances 1, 2 and 3, likewise, are the same as the end group resonances observed for a linear polyethylene. Thus a six carbon branch produces the same C-13 spectral pattern as any subsequent branch of greater length. Carbon-13 NMR, alone, therefore cannot be used to distinguish a linear six carbon branch from a branch of some intermediate length or a true long chain branch.

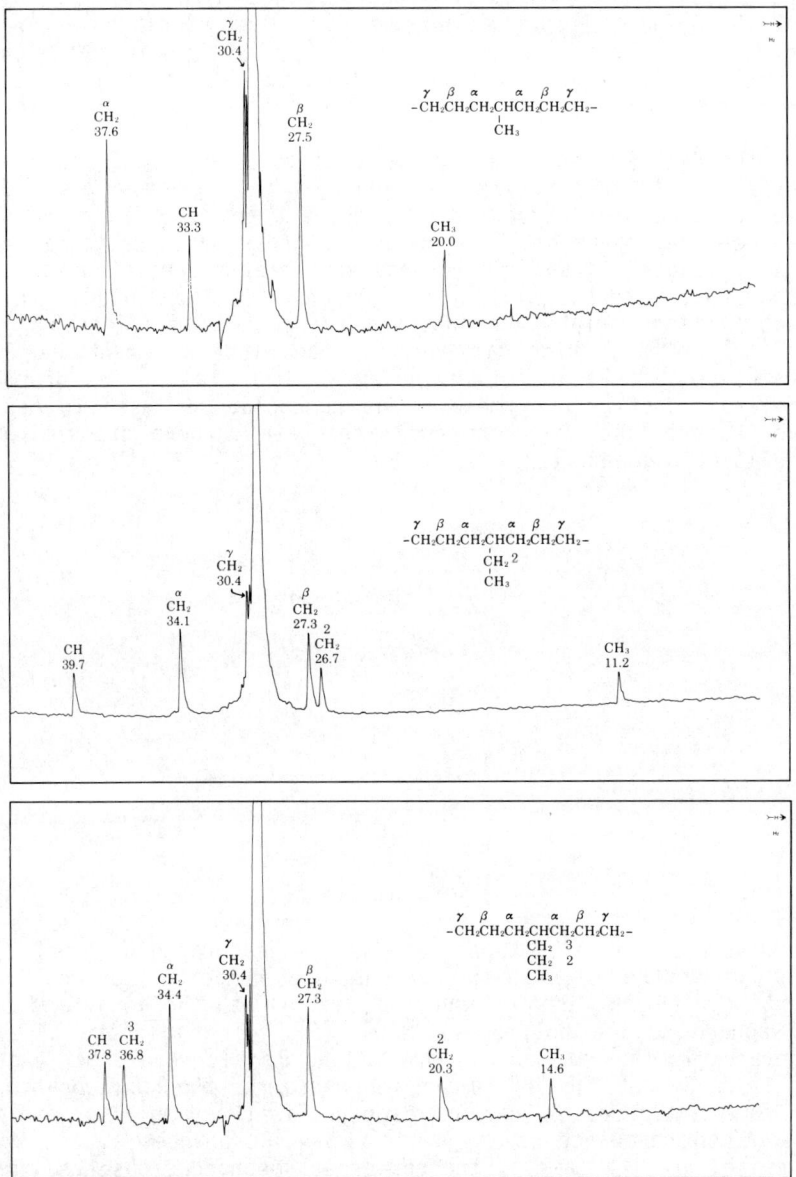

Figure 2. C-13 NMR spectra at 25.2 MHz of (top) *an ethylene-1-propene copolymer,* (middle) *an ethylene-1-butene copolymer, and* (bottom) *an ethylene-1-pentene copolymer*

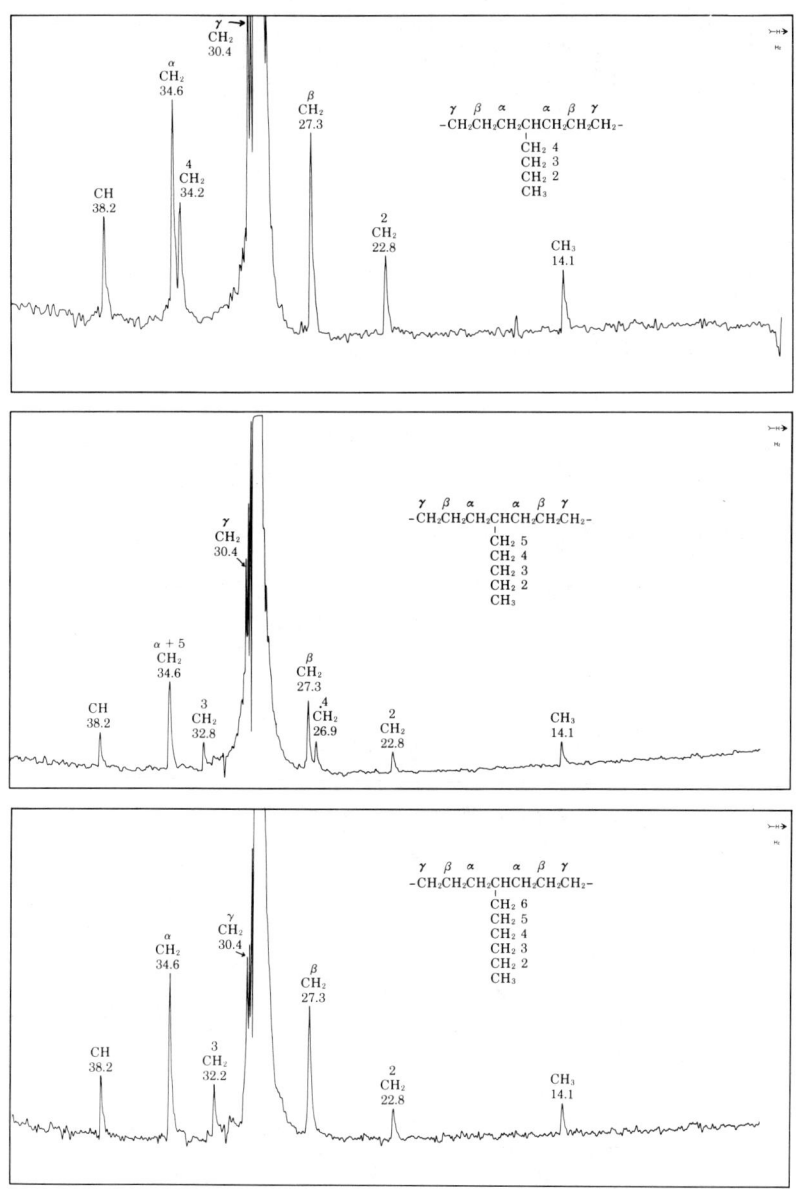

Figure 3. C-13 NMR spectra at 25.2 MHz of (top) *an ethylene-1-hexene copolymer,* (middle) *an ethylene-1-heptene copolymer, and* (bottom) *an ethylene-1-octene copolymer*

TABLE I

Polyethylene Backbone and Side-Chain C-13 Chemical Shifts in ppm from TMS (± 0.1) as a Function of Branch Length (γ Carbon Chemical Shifts, which occur near 30.4 ppm, are not given because they are often obscured by the major 30 ppm resonance for the "n" equivalent, recurring methylene carbons). Solvent: 1,2,4-trichlorobenzene. Temperature: 125°C.

Branch Length	Methine	α	β	1	2	3	4	5	6
1	33.3	37.6	27.5	20.0					
2	39.7	34.1	27.3	11.2	26.7				
3	37.8	34.4	27.3	14.6	20.3	36.8			
4	38.2	34.6	27.3	14.1	23.4	-----	34.2		
5	38.2	34.6	27.3	14.1	22.8	32.8	26.9	34.6	
6	38.2	34.6	27.3	14.1	22.8	32.2	30.4	27.3	34.6

6. RANDALL *Long-Chain Branching in Polyethylenes* 101

The capability for discerning the length of short chain branches has made C-13 NMR a powerful tool for characterizing low density polyethylenes produced from free radical, high pressure processes. The C-13 NMR spectrum of such a polyethylene is shown in Figure 4. It is evident that the major short chain branches are butyl, amyl and ethyl. Others are also present, and Axelson, Mandelkern, and Levy, in a comprehensive study (6) have concluded that no unique structure can be used to characterize low density polyethylenes. They have found nonlinear short chain branches as well as 1,3 paired ethyl branches. Bovey, Schilling, McCracken and Wagner (9) compared the content of branches six and longer in low density polyethylenes with the long chain branching results obtained through a combination of gel permeation chromatography and intrinsic viscosity. An observed good agreement led to the conclusion that the principal short chain branches contained fewer than six carbons and the six and longer branching content could be related entirely to long chain branching. Others have now reported similar observations in studies where solution methods are combined with C-13 NMR (10). However, as a result of the possible uncertainty of the branch lengths, associated with the resonances for branches six carbons and longer, C-13 NMR should be used in conjunction with independent methods to establish true long chain branching.

From the results we have seen thus far, it is easy to predict the C-13 NMR spectrum anticipated for essentially linear polyethylenes containing a small degree of long chain branching. An examination of a C-13 NMR spectrum from a completely linear polyethylene, containing both terminal olefinic and saturated end groups, shows that only five resonances are produced. A major resonance at 30 ppm arises from equivalent, recurring methylene carbons, designated as "n", which are four or more removed from an end group or a branch. Resonances at 14.1, 22.9 and 32.3 ppm are from carbons 1,2 and 3, respectively, from the saturated, linear end group. A final resonance which is observed at 33.9 ppm, arises from an allylic carbon, designated as "a", from a terminal olefinic end group. These resonances, depicted structurally below, are fundamental to the spectra of all polyethylenes.

$$CH_3-CH_2-CH_2-CH_2-CH_2- \quad\quad -(CH_2)_n- \quad\quad -CH_2-CH=CH_2$$
$$1 \quad 2 \quad 3 \quad\quad\quad\quad\quad "n" \quad\quad\quad\quad "a"$$

An introduction of branching, either long or short, will create additional resonances to those described above. For long chain branches, these will be an α, β, (and sometimes γ) and a methine resonance as depicted structurally below:

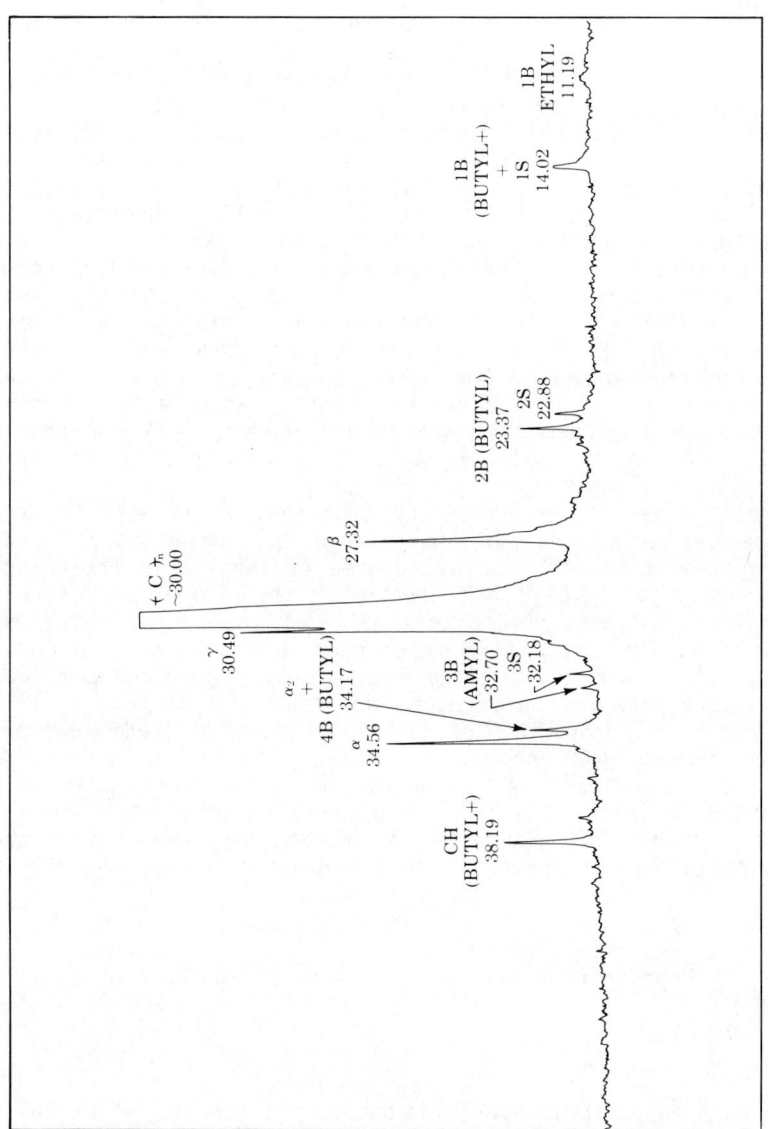

Figure 4. C-13 NMR spectrum at 25.2 MHz of a low-density PE produced from a high-pressure process

$$\begin{array}{c}\text{[γ]}\quad\beta\quad\alpha\qquad\alpha\quad\beta\quad\text{[γ]}\\\text{-CH}_2\text{-CH}_2\text{-CH}_2\text{-CH}_2\text{-CH-CH}_2\text{-CH}_2\text{-CH}_2\text{-CH}_2\text{-}\\\alpha\quad|\\\text{CH}_2\\\beta\quad\text{CH}_2\\\text{[γ]}\quad\text{CH}_2\\\phantom{\text{[γ]}\quad}\text{CH}_2\end{array}$$

From the observed C-13 NMR spectrum of the ethylene-1-octene copolymer (Figure 3), we find that the α, β and methine resonances associated with branches six carbons and longer occur at 34.56, 27.32 and 38.17 ppm, respectively. Thus in high density polyethylenes, where long chain branching is essentially the only type present, carbon-13 NMR can be used to establish unequivocally the presence of branches six carbons long and longer. If no comonomer has been used during polymerization, it is very likely that the presence of such resonances will be indicative of true long chain branching. In any event, C-13 NMR can be used to pinpoint the absence of long chain branching and place an upper limit upon the long chain branch concentration whenever branches six carbons and longer are detected.

The need for a complementary measurement to C-13 NMR in studies of long chain branching should be apparent. It has been pointed out that a promising possibility appears to be flow activation energies obtained from dynamic shear moduli as a function of temperature (20). Flow activation energies range from approximately 6.0 kcal/mol for linear systems to around 13.5 kcal/mol for systems containing extensive long chain branching. Chain entanglements are one of the factors influencing flow activation energies, but to the extent that long chain branches also influence chain entanglements, this technique can be an indicator of the presence of long chain branching. Four polyethylenes, labelled "A" through "D" and selected for C-13 NMR characterization on a basis of the observed flow activation energies, are described in Table II. A fifth polymer, called "E", was also examined as a reference polymer because it was not expected to contain any significant long chain branching as indicated by its flow activation energy (see Table II). Carbon-13 NMR data were obtained at a high field (50 MHz, 47 kG) to achieve improved sensitivity. The 50 MHz spectra of these polymers, A through E, are reproduced in Figures 5 through 9. Instrumental conditions necessary for quantitative measurements will be discussed later; for the present, we will be concerned with the use of C-13 NMR as a means for simply detecting the presence of long chain branching.

TABLE II

Molecular Weights, Flow Activation Energies and Type of Catalyst for a Series of Polyethylenes Examined for Long Chain Branching.

Polymer	M_w	M_n	E_a	Catalyst	Comonomer	"g" factor*
A	159,000	19,100	8.0 kcal/mol	chromium based	None	0.8
B	224,000	13,600	8.6 kcal/mol	chromium based	None	0.5
C	148,000	17,900	9.3 kcal/mol	chromium based	None	0.6
D	226,000	8,500	9.6 kcal/mol	chromium based	Hexene-1	0.7
E	172,000	32,400	6.1 kcal/mol	titanium based	None	(1+)

*These "g" factors are probably not very accurate. See text.

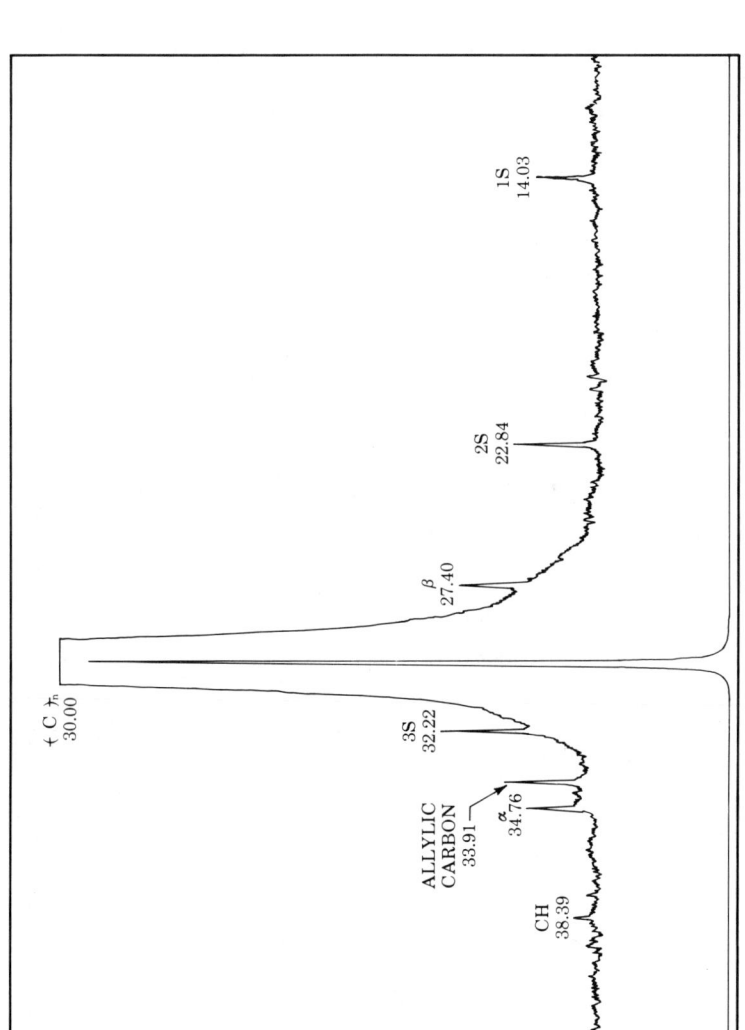

Figure 5. C-13 NMR spectrum at 50 MHz of Polymer A (~ 75% in trichlorobenzene at 125°C, number of transients accumulated 45,300). Spectrum was provided courtesy of Nicolet Technology Corporation.

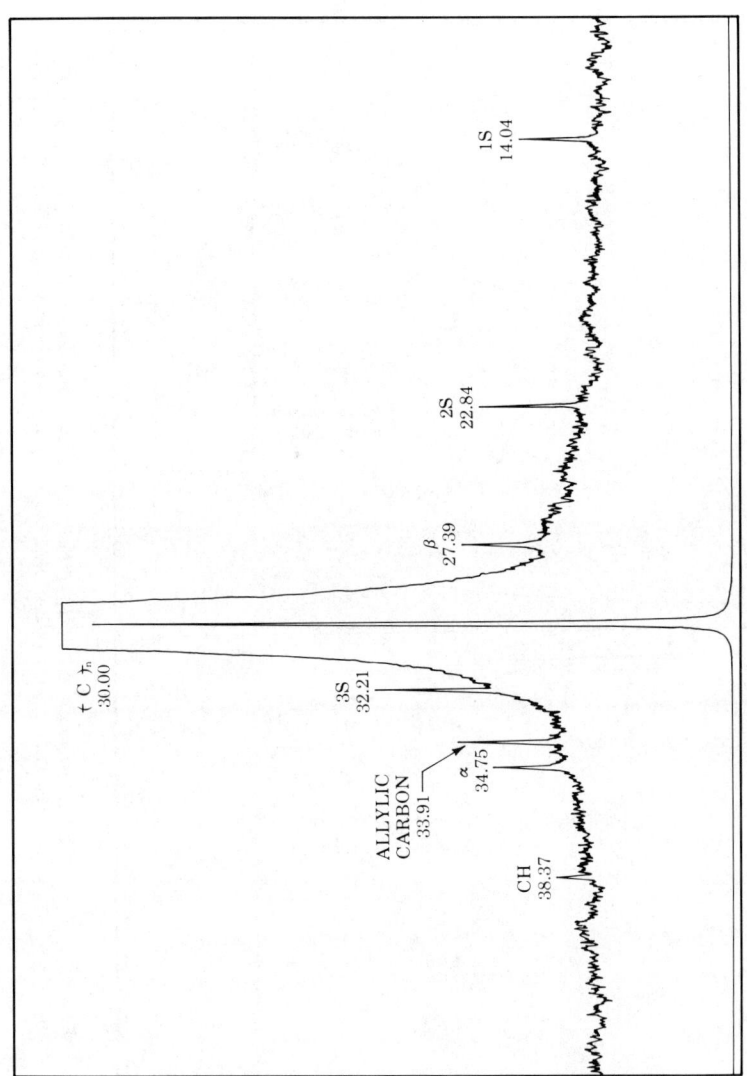

Figure 6. C-13 NMR spectrum at 50 MHz of Polymer B (∼ 75% in trichlorobenzene at 125°C, number of transients accumulated 4,900). Spectrum was provided courtesy of Nicolet Technology Corporation.

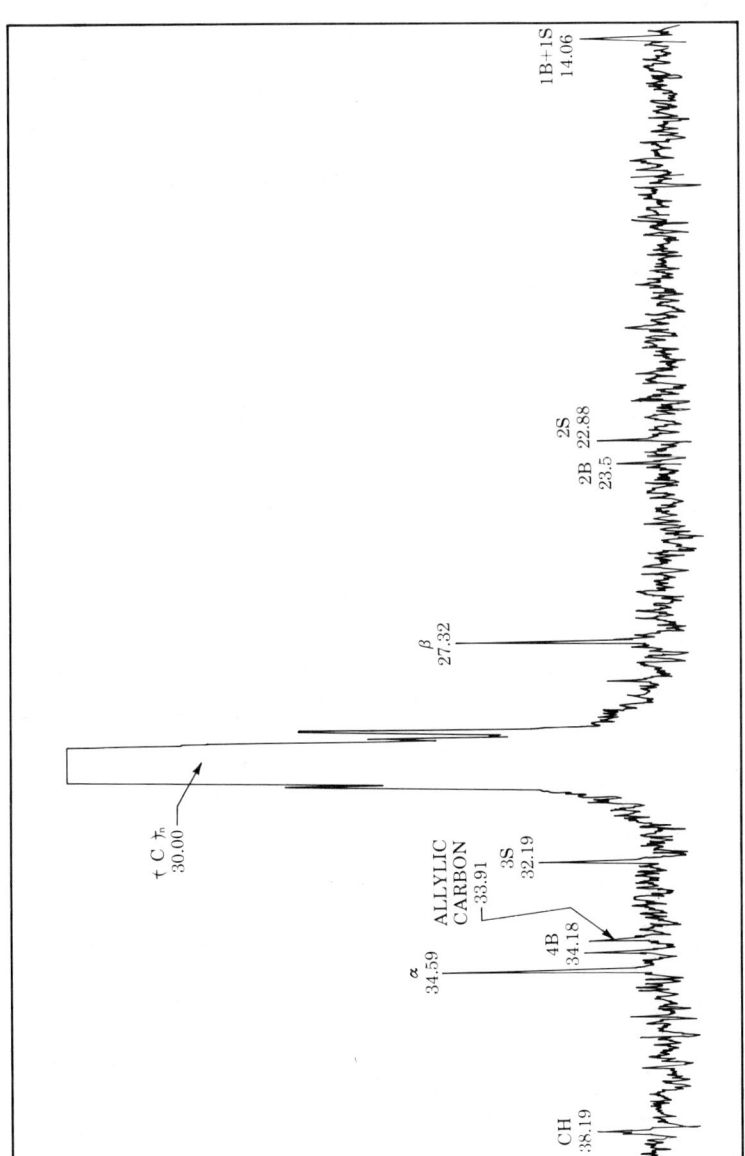

Figure 7. C-13 NMR spectrum at 50 MHz of Polymer C PE homopolymer (~ 10% in trichlorobenzene at 125°C, number of transients accumulated 50,000). Spectrum was provided courtesy of Varian Associates.

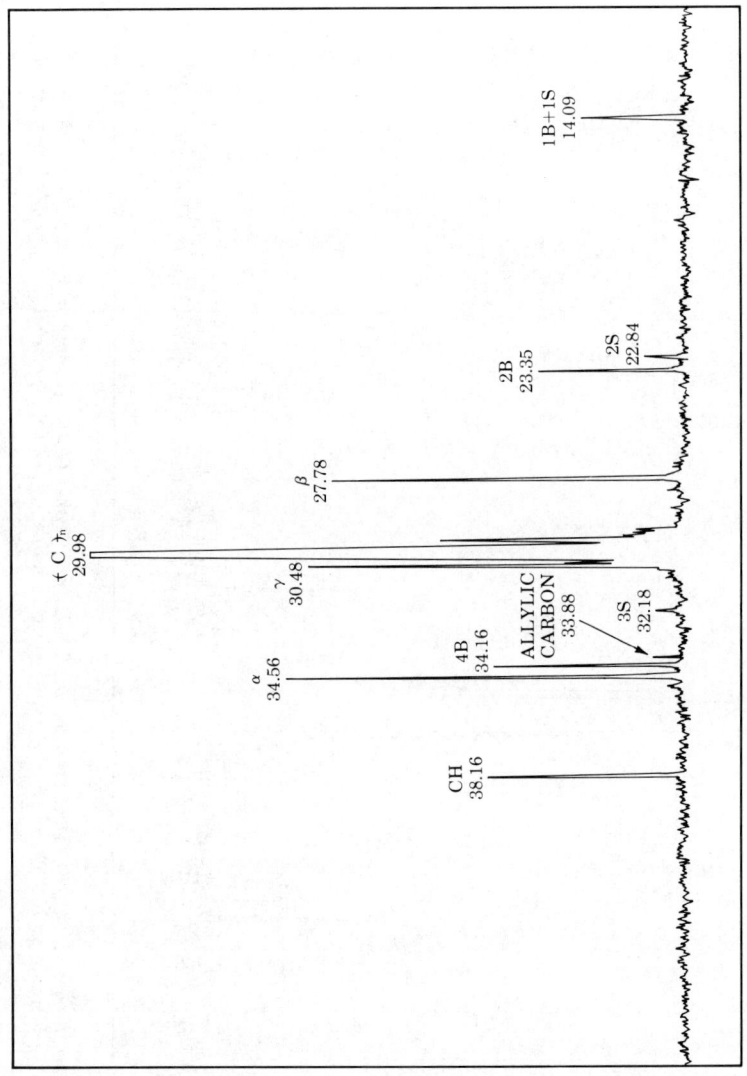

Figure 8. C-13 NMR spectrum at 50 MHz of Polymer D ethylene-1-hexene copolymer (~ 10% in trichlorobenzene at 125°C, number of transients accumulated 8,743). Spectrum was provided courtesy of Varian Associates.

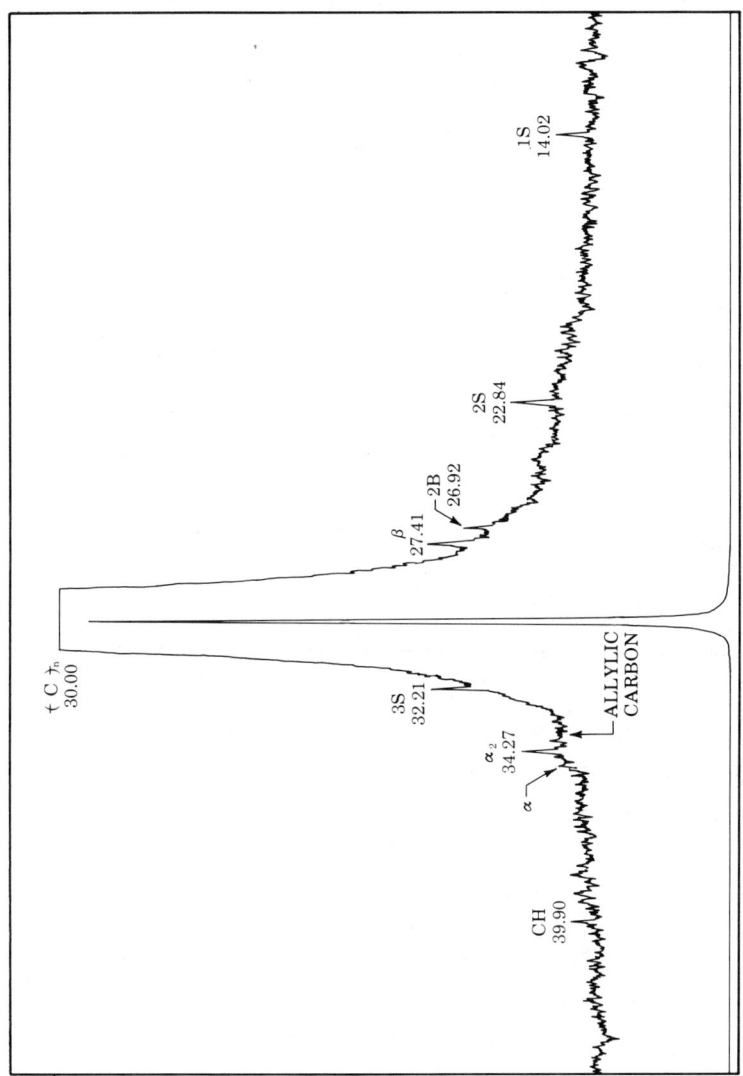

Figure 9. C-13 NMR spectrum at 50 MHz of Polymer E (~75% in trichlorobenzene at 125°C, number of transients accumulated 10,100). Spectrum was provided courtesy of Nicolet Technology Corporation.

For those polyethylenes where short chain branching is present in addition to long chain branching (see Figures 8 and 9), a modification in the nomenclature scheme is needed to distinguish identifiable short chain carbon resonances from the corresponding carbons in chain end groups. Consequently, an "S" (for saturated end groups) has been added to the 1, 2 and 3 carbon nomenclature for chain ends. The numbers for carbons in branches of length five and shorter have a "B" added. In figures 5-9, the resonances from chain ends, six and longer, are clearly identified from carbon resonances from the short side-chain branches.

The 50 MHz C-13 NMR spectrum of polymer "A" is reproduced in Figure 5. It is nearly a classical representation of the spectrum anticipated for a polyethylene containing long chain branching. Only the five resonances expected for linear polyethylene systems plus the α, β, and methine resonances for long chain branches are observed. A simple inspection of the α, β intensities as compared to the end group resonances indicates that fewer than one polymer molecule out of three has a long chain branch. A similar result was obtained for polymer "B" shown in Figure 6. Both polymers "A" and "B" contain long chain branches and, on an average, have approximately equal numbers of both saturated and terminal olefin end groups.

Polymer "C", shown in Figure 7, gives a more complex C-13 NMR spectrum than observed for polymers "A" and "B" because four-carbon side-chain branches are positively indicated even though no comonomer was added during polymerization. The presence of butyl branches complicates an identification of long chain branching because the α, β resonances from butyl branches have the same chemical shifts as do the α, β resonances from longer, linear branches (see Table I). The long chain branching for systems containing butyl branches, therefore, can only be determined from the differences observed in the relative intensities for the "4" carbon and the α carbon. In Figure 3, the α to "4" carbon resonance intensities are distinctly 2:1. In Figure 7, the relative intensities for the α to "4" carbon resonances are 2.7:1 and long chain branching is, therefore, indicated. The polymerization conditions have somehow led to an introduction of butyl branches without also introducing amyl branches or branch lengths shorter than four. This result is interesting and leads to the suggestion that the long chain branches in polymer "C" may not be truly "long". A disproportionate number of these "long chain" branches may be intermediate in length. Further study is clearly indicated for this polyethylene system. Without C-13 NMR, however, one may not have been aware that short chain branches were being introduced into a polyethylene polymerization without adding comonomer.

Polymer "D" is an ethylene-1-hexene copolymer by design. A flow activation energy of 9.6 kcal/mol suggests that long chain branching may be present. The C-13 NMR spectrum, however, is complicated by the presence of butyl branches as in the case of polymer "C". Once again, the detection of long chain branching must be based on a ratio of the observed intensities for the "α" carbon resonance to the "4" carbon resonance, which is 2.04 in this case and within experimental error of the NMR intensity measurements. Thus long chain branching cannot be determined readily unless one resorts to a tedious set of measurements designed to establish whether the excess intensity of the "α" resonance is outside the limits of experimental error.

Polymer "E" was prepared with a Ziegler type of catalyst and has only a small quantity of terminal olefin end groups, as shown by the NMR spectrum in Figure 9. It was not expected to have any significant degree of long chain branching because the flow activation energy was 6.1 kcal/mol. The C-13 NMR spectrum is consistent with the presence of very low quantities of long chain branching (see "α" in Figure 9). An unexpected result did occur, however, because ethyl branches could be clearly identified by the chemical shifts for the "α_2", "β", methine and side-chain carbon resonances. Note in Figure 9 that the methyl resonance for the ethyl branch is not shown; the others can be clearly identified in Figure 9).

From a quantitative viewpoint, it is evident that the relative intensities of the resonances from carbons associated with branches and end groups can be compared to the intensity for the major methylene resonance, "n", at 30.00 ppm to determine branch concentrations and number average molecular weight or carbon number. The following definitions are useful in formulating the appropriate algebraic relationships:

n = intensity of the major methylene resonance at 30 ppm

$\bar{s}$ = average intensity for a saturated end group carbon

a = the allylic carbon intensity at 33.9 ppm

C_{tot} = the total carbon intensity

$\bar{\alpha}$ = 1/2 (α + β) carbon intensities

N = average number of long chain branches per polymer molecule

$N+2$ = average number of end groups per polymer molecule

With the previous definitions, the polymer carbon number and number average molecular weight are given by:

$$\text{Carbon Number} = C_{tot}(N+2)/(\bar{s} + a) \quad (1)$$

$$M_n = 14 \times \text{Carbon Number} \quad (2)$$

For linear polymers where "N" is zero, the carbon number and number average molecular weight can be easily and reliably determined. For those polymers containing long chain branching, one must utilize the ratio of the carbon resonance intensities associated with branching to the end group carbon resonance intensities to determine "N" as follows:

$$N = 2\bar{\alpha}/(3(\bar{s} + a) - \bar{\alpha}) \quad (3)$$

The number of long chain branches per ten thousand carbon atoms is similarly given by:

$$\text{Branches}/10,000 \; C = ((1/3 \; \bar{\alpha})/(C_{tot} \times 10^4)) \quad (4)$$

Equation 4 can be easily modified for the number of short chain branches (fewer than six carbons) per 10,000 carbons by replacing $1/3 \; \alpha$ with an intensity appropriate for one branch carbon from the short chain branch.

With the comprehensive structural information available from C-13 NMR analyses of both low and high density polyethylenes, quantitative measurements become highly desirable. As seen from equations 1-4, the quantitative relationships necessary for analyses of branching, molecular weight and end group distributions can be readily derived without resorting to limiting assumptions. The assignments have been rigorously established; thus the only question remaining is the reliability of the data with respect to the observed relative intensities. Differences among nuclear Overhauser effects, non-equilibrium dynamic conditions during data gathering and software-hardware problems with the dynamic range are factors which would adversely affect the relative observed C-13 NMR spectral intensities.

It has been demonstrated by Levy and coworkers (6)(21) that nuclear Overhauser effects in polyethylenes are full and, therefore, are not a consideration under the experimental conditions (~10% solutions, 125°C) employed in C-13 NMR quantitative measurements of polyethylene. Because the C-13 NMR experiment utilizes the Fourier transform technique and free induction decay data is gathered in a time dependent framework, spin-lattice relaxation times, T_1's, are important factors that must be considered when designing a pulse sequence for long term data averaging. Generally, the pulse spacings must be five times the longest observed T_1 to ensure complete relaxation between rf pulses (22). The methyl groups in chain ends and short branches can have T_1's ranging from 3 to 7 seconds (21); thus the time required to gather three to five thousand FID's can become inordinately long. When planning quantitative experiments on well-known systems, it is advisable to select those resonances most favorable from a T_1 standpoint for quantitative measurements because instrument time is expensive and should be used efficiently.

Dynamic range is an important consideration in quantitative NMR measurements because small resonances may become truncated with respect to large resonances ($\geq$1000:1) during time averaging if a sufficient computer word length is not available (23). It is imperative in measurements of long chain branching in polyethylene that an NMR instrument be equipped with either a 16K computer system with block averaging or have a 16K computer system with either double precision or floating point arithmetic. Instruments are now available that have both double precision and floating point arithmetic, which are also valuable in preventing unwanted truncations during Fourier transformations. With reasonable care given to both the software and hardware problems associated with quantitative C-13 NMR of polymers, one finds that measurements of end group distributions, molecular weight, and branching in polethylene are routinely available.

One of the best tests for satisfactory NMR instrumental conditions during quantitative C-13 NMR measurements is to examine known reference standards. The best for this purpose are NBS standards 1482 (M_w = 13,600, M_n = 11,400) and 1483 (M_w = 32,100, M_n = 28,900). The NBS standard 1483 was examined under precisely the same experimental NMR conditions as polyethylenes C and E reported in Table II. The C-13 NMR spectrum is shown in Figure 10. Instrumental conditions and pertinent intensity data are given below:

	Resonance	Peak Height	Assignment
Pulse Angle: 50°	32.18 ppm	4.5	3S
Pulse Spacing: 1 second	33.91 ppm	1.1	"a"
Double Precision Arithmetic	30.00 ppm	6082	"n"
	34.09 ppm	3.9	α_2

M_n = 30,560 (Calculated from Equations 1 and 2).

(Only 3S was used to determine $\bar{s}$ because it has the shortest T_1 of the three terminal carbons.)

As was the case for polymer "E", a small amount of ethyl branching was detected in NBS 1483 as indicated in Figure 10 by the resonances observed for the appropriate methine, α_2, β and 2B carbons. (The methyl resonance is not shown.) The concentration of ethyl branches is 3 per 10,000 carbons as determined using a modified version of Equation 4. NBS 1483 was probably prepared using a Ziegler type catalyst system, as suggested by the rela-

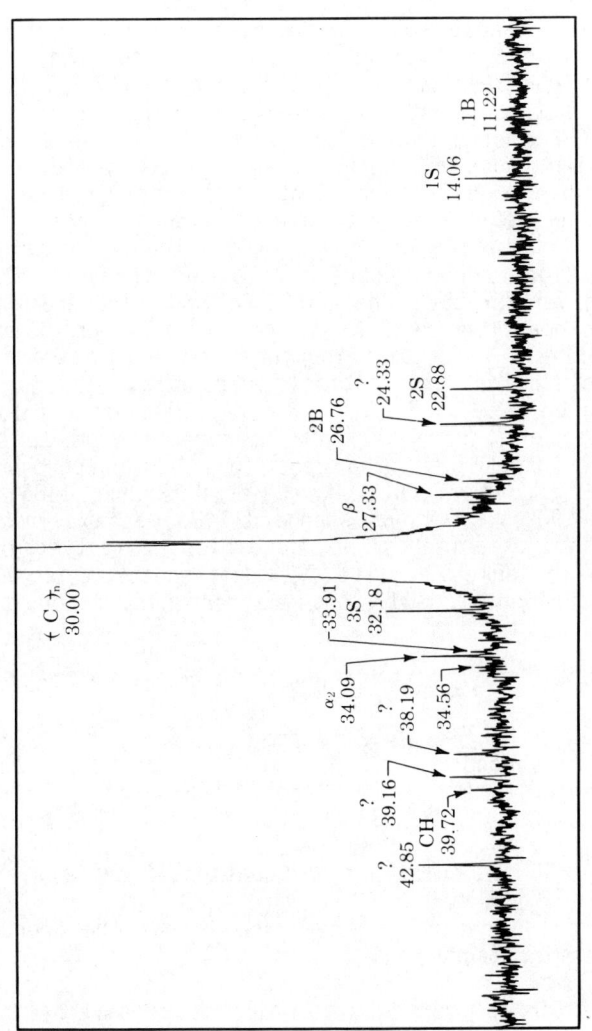

Figure 10. C-13 NMR spectrum at 50 MHz of NBS 1483 (~10% in trichlorobenzene at 125°C, number of transients accumulated 229,462). Spectrum was provided courtesy of Varian Associates.

tively low intensity of the 33.9 resonance for the allylic carbon. It is interesting that a low amount of ethyl branching was detected in this sample and in polymer "E", which was prepared with a Ziegler type catalyst. NBS 1482 was examined using an XL-100 spectrometer at 25 MHz. Instrumental conditions and intensity data are given below:

	Resonance	Peak Height	Assignment
Pulse Angle: 90°	14.09 ppm	11	1S
Pulse Spacing: 10 seconds	22.86 ppm	15	2S
Double Precision Arithmetic	32.16 ppm	14	3S
	33.00 ppm	8	"a"
	29.98 ppm	9574	n

$\bar{s}$ = 14.5 (1S was not used).

M_n = 11,970 (Calculated from Equations 1 and 2).

Ethyl branching was not detected in NBS 1482; however, the instrumental sensitivity was such that branching in a range of 1-5 per 10,000 carbons would not be observed. The agreement between the number average molecular weights calculated from the NMR data and that reported by NBS serves to indicate the viability of the NMR method for determining number average molecular weight. The method discussed above, when applied to the C-13 NMR data from polymers A through E, gave the results listed in Table III for M_n, N and degree of branching.

TABLE III

Number Average Molecular Weight, Long Chain Branching, Short Chain Branching and End Group Distribution for Polymers A, B, C, D and E.

Polymer	C_{6+} Branches/ Molecule	Branches/10,000		Mn	s/a
A	0.28	(C_{6+})	1.8	21,700	1/1
B	0.29	(C_{6+})	2.1	18,680	1/1
C	0.23	(C_{6+})	1.4)	23,100	1/7.1
		(Butyl)	5.5)		
D	----	(Butyl)	64	10,300	1/1
E	----	(Ethyl)	2.2	28,650	2.2/1

Number average molecular weight data from size exclusion chromatography for polymers A through E has been given previously in Table II. These results should be compared to the number average molecular weights obtained from NMR in Table III. In size exclusion chromatography, both linear and branched molecules are separated according to their respective hydrodynamic volumes, that is

$$n_{lin}M_{lin} = n_{br}M_{br}, \quad (5)$$

the principal of universal calibration (18). In this particular SEC analysis, no consideration was given to the possibility that the polymer molecules could be branched; thus low molecular weight results should be anticipated because branched polymers have overall smaller dimensions than linear polymers of the same molecular weight. Lower results were obtained as shown by the data in Table II although the internal trends are identical to those obtained from NMR. It is possible to use the number average molecular weights from SEC and NMR to calculate the familiar "g" factor (18) through the relationship,

$$g^{1/2} = n_{br}/n_{lin} \quad (6)$$

and equation 5. Values of "g" less than one were obtained although they are somewhat smaller than the values predicted by Zimm and Stockmayer (24) for the amount of long chain branching determined by NMR. The differences could be easily accounted by the error ($\pm$ 10% in the molecular weights from SEC). Although these "g" factors are probably inaccurate, they are in the correct direction from the NMR data and from SEC.

Polymer "E" gave the highest number average molecular weight of the polymers examined, and for this reason, it probably gave the least accurate result from both NMR and SEC, which are probably within experimental error. The error in the NMR measurement will increase with molecular weight unless an effort is made to obtain spectra with the same signal-to-noise ratio for the respective resonances associated with branches. The utility and accuracy of the C-13 NMR method in providing quantitative polymer structural data, however, is gratifying.

The C-13 NMR method has been criticized in the past because it is time consuming. Five to ten thousand transients with 10-15 seconds pulse delays are usually required with 10-15% by weight solutions to obtain signal-to-noise adequate for a one part in one thousand measurement. This time factor can be reduced substantially if one uses twenty millimeter sample tubes and a superconducting magnet system and examines the polyethylene in a melt state. This improvement plus the fact that three measurements, number average molecular weight, end group distribution and degree of branching, are accomplished in one make C-13 NMR a highly attractive method for characterizing polyethylenes. A serious drawback is not encountered even though branches six carbons in length and longer are measured collectively. The short branches are generally less than six carbons in length and truly long chain branches tend to predominate. On occasions, there may be special exceptions for "intermediate" branch lengths, as shown by polymer "C" in this study, so independent rheological measurements should be sought as a matter of course. Nevertheless, a direct method, which possesses the required sensitivity to determine long chain branching in high density polyethylenes, is now available.

REFERENCES

1. M. L. Miller, "The Structure of Polymers", Reinhold, New York, 1966 p. 118.

2. A. H. Willbourn, J. Polym. Sci., 34, 569 (1959) Nottingham Symposium.

3. E. P. Otocka, R. J. Roe, M. Y. Hellman, and P. M. Muglia, Macromolecules, 4, 507 (1971).

4. D. E. Dorman, E. P. Otocka, and F. A. Bovey, Macromolecules, 5, 574 (1972).

5. J. C. Randall, J. Polym. Sci., Polym. Phys. Ed., 11, 275 (1973).

6. D. E. Axelson, G. C. Levy and L. Mandelkern, Macromolecules, 12, 41 (1979).

7. M. E. A. Cudby and A. Bunn, Polymer, 17, 345 (1976).

8. J. C. Randall, J. Appl. Polym. Sci., 22, 585 (1978).

9. F. A. Bovey, F. C. Schilling, F. L. McCrackin and H. L. Wagner, Macromolecules, 9, 76 (1976).

10. G. N. Foster, Polymer Preprints, 20, 463, (1979).

11. D. E. Axelson and W. C. Knapp (in press).

12. J. P. Hogan, C. T. Levett and R. T. Werkman, S. P. E. Journal 23(11), 87 (1967).

13. G. Kraus and C. J. Stacy, J. Polym. Sci., Symposium No. 43, 329 (1973).

14. S. D. Abbott, American Laboratory, August, 1977, p. 41.

15. D. R. Rueda, F. J. Balta Calleja and A. Hidalgo, Spectrochimica Acta, 30A, 1545 (1974).

16. D. R. Rueda, F. J. Balta Calleja and A. Hidalgo, Spectrochimica Acta, 35A, 847 (1979).

17. A. Barlow, L. Wild, and R. Ranganath, J. Appl. Polym. Sci., 21, 3319 (1977).

18. L. Wild, R. Ranganath and A. Barlow, J. Appl. Polym. Sci, 21, 3331 (1977).

19. D. M. Grant and E. G. Paul, J. Am. Chem. Soc., 86, 2984 (1964).

20. J. K. Hughes, submitted for publication.

21. D. E. Axelson, L. Mandelkern, and G. C. Levy, Macromolecules, 10, 557 (1977).

22. T. C. Farrar and E. D. Becker, "Pulse and Fourier Transform NMR", Academic Press, New York (1971) p. 21.

23. J. C. Randall, "Polymer Sequence Determination: Carbon-13 NMR Method", Academic Press, New York (1977) p. 100.

24. B. H. Zimm and W. H. Stockmayer, J. Chem. Phys., 17, 130 (1949).

RECEIVED June 20, 1980.

Molecular Dynamics of Polymer Chains and Alkyl Groups in Solution

GEORGE C. LEVY, PETER L. RINALDI, JAMES J. DECHTER, DAVID E. AXELSON, and LEO MANDELKERN

Department of Chemistry, The Florida State University, Tallahassee, FL 32306

For carbons possessing a directly bonded proton, the carbon-13 spin lattice relaxation time, T_1, spin-spin relaxation time, T_2, and the nuclear Overhauser enhancement, NOE, comprise a set of parameters which characterize molecular motions. In the case of simple isotropic motion, the dependence is in terms of a single correlation time characterizing the exponential decay of the autocorrelation function. However, in many instances, the assumption of isotropic motion is not valid. For rigid systems, the relaxation behavior can then often be predicted by assuming simple anisotropic motion (1). Often, superposition of two or more independent motions must be used to satisfactorily interpret observed relaxation behavior (1,2). Recently, however, the widespread availability of ^{13}C nmr instruments, has led to a number of examples where these models have proved unsatisfactory.

Deviations from predicted relaxation behavior have been observed for large proteins (3-7), polymers (8,9) and highly associated small molecules (10). Particularly prominent are observations of T_1 field dependences and low NOE's within the so-called "extreme spectral narrowing region," where single correlation time models predict field independence of T_1 and full NOE's.

A number of theories have been invoked to describe molecular motion consistent with observed relaxation behavior, including: (1) anisotropic rotational diffusion (11), (2) use of distributions of correlation times (8,9,12,13,14), (3) restricted internal diffusion (15,16) and (4) librational motion (17). In one instance the incorrect estimation of C-H bond distances was shown to affect calculations of motional correlation times (18), particularly when $\omega^2\tau_c^2$ approaches unity. C-H bond length variations, however, do not explain the anomalies in a majority of cases, especially since NOE's, which are insensitive to C-H distance, are not correctly predicted (vide infra).

We have been interested in studying molecular dynamics of polymer chains and alkyl groups anchored at one end, particularly n-alkyl glycol and glycerol derivatives, poly(n-butyl acrylate) (PBA), poly(n-butyl methacrylate) (PBMA), and poly(n-hexyl meth-

0-8412-0594-9/80/47-142-119$06.75/0
© 1980 American Chemical Society

acrylate)(PHMA).

When single frequency ^{13}C relaxation data are applied to these systems, the various theoretical models are more or less indistinguishable. The approach used in our laboratory has been to measure relaxation parameters at a minimum of two widely separated magnetic fields (8,10). Under these conditions, it is possible to more closely describe the complex dynamics of these systems.

Howarth(17b) has used the theory of Internal Librational Motion to successfully predict the field dependent relaxation behavior of the 1,2-decanediol (DD), PBMA, and PHMA systems (using our published experimental data). We have utilized together multiple internal rotations (MIR) and distributions of correlation times. These methods individually have been successful in predicting relaxation behavior at one field. However, only the distribution theory predicts the observed field dependence for the carbons at or near sites of motional restriction, yet still having apparent correlation times $\lesssim 10^{-10}$ sec. Our interest in the study of concerted motions along these alkyl chains has led us to combine the two approaches in the treatment of ^{13}C relaxation parameters.

Experimental

1,2-Decanediol (DD) and 1,2-hexadecanediol (HDD) were synthesized from their respective alkenes(19). 1,2,3-decanetriols (DT) were prepared by a procedure described in the literature (20). Poly(n-butyl methacrylate)(PBMA) was obtained from Polyscience as high molecular weight material. Solutions were made without further purification using toluene-d_8 as a solvent. Poly(n-hexyl methacrylate)(PHMA) and Poly(n-butyl acrylate)(PBA) were purchased as toluene solutions (25 wt%) from Aldrich Chemical Co. PBA and PHMA samples (50% w/w) were prepared by solvent evaporation at ~70° under a N_2 gas stream.

Natural abundance ^{13}C spectra were obtained using quadrature detection modified Bruker HX-270 and HFX-90 spectrometers operating at 67.9 and 22.6 MHz, respectively. Free induction decays were accumulated using 4K/4K data points and a ±3kHz spectral window. A fast inversion recovery (FIRFT) pulse sequence (21) was employed to measure T_1's; T_1's were calculated using a nonlinear three parameter fitting procedure(22).

Nuclear Overhauser enhancement factors (NOEF's) were determined using gated decoupling. Two sets of two spectra were obtained, alternately with two level(23) continuous wideband decoupling and gated decoupling, with a pulse interval greater than ten times the longest T_1. NOEF values are taken from the average of the two data sets and are accurate to better than ±15%. Temperature was controlled using a Bruker BST-100 heating unit which was calibrated by measuring the temperature of a tube containing ethylene glycol and a thermometer, placed in the probe.

Samples were equilibrated in the probe for at least 20 minutes before spectra were obtained.

Results and Discussion

Poly(n-butyl methacrylate). The ^{13}C T_1's and NOEF's of PBMA are presented in Tables I and II, respectively. At a given temperature the T_1's increase with distance from the motional restrictions imposed by the chain backbone. The main chain CH_2 group and the first protonated side chain carbon C-1 both show this restriction of motion. The T_1's of C-1 are relatively constant with temperature considering the large temperature range (~100°) over which they were examined. The apparent minima for C-1 are at 0.28 and 0.09 sec at 67.9 and 22.6 MHz, respectively. This is significantly larger than the values of 0.055 and 0.018 sec predicted by the single correlation time model. Similar results are also observed for C-CH_2, the polymer backbone methylene carbon. Most important of all, is the observed field dependence of all the ^{13}C T_1's over the entire temperature range. This is surprising, since C-2, C-3 and especially C-4 have T_1's which correspond to correlation times well within the extreme spectral narrowing condition $[(\omega_H+\omega_C)^2\tau_c^2<<1]$ as normally applied, where dipolar T_1's should be field-independent. This, of course, is equivalent to stating that while the average motion may be rapid, components of this motion are correlated with longer time scale dynamics. The observation of field-dependent T_1's exceeding 0.2 sec for the side chain carbons indicates that the effective autocorrelation function decays non-exponentially and thus the spectral density function is frequency dependent over an extended range of molecular dynamics.

The NOEF behavior of the carbon resonances in PBMA is also surprising (Table II). For C-1 at 67 MHz and low temperature, precisely where the NOEF is predicted to be smallest, it is maximum. Contrary to simple prediction, the NOEF's decrease with increasing temperature. At low field (22 MHz) the opposite behavior is observed. C-2 and C-4 exhibit qualitatively similar behavior at 67 MHz, but there is no definite pattern to the NOEF's at 22 MHz except that they are all significantly less than the theoretical maximum of 1.99.

While use of multiple internal rotations does allow prediction of the T_1's at any one field better than a single correlation time theory, it still does not predict a T_1 field dependence. Thus, this better fit may be solely due to the introduction of additional variables, however carefully utilized. The predicted NOEF behavior of C-1 undergoing internal rotation superimposed upon overall isotropic motion is shown in Figure 1. Qualitatively, behavior such as that observed in PBMA can be interpreted in terms of crossovers arising from the mixing of overall motional components and rapid internal motional modes. The resulting NOEF values show complex field dependences.

Table I. ^{13}C Spin Lattice Relaxation Times of Poly(n-butyl methacrylate) in 50% (w/w) Solution in Toluene-d_8

Temp. °C	\multicolumn{7}{c}{T_1 (67.9MHz), s}						
	C-1	C-2	C-3	C-4	CCH_3	CCH_2	CCR_4
-5	0.27	0.20	0.35	0.93	0.053	a	2.7
6	0.35	0.23	0.39	1.0	0.057	0.21	2.3
21	0.35	0.37	0.73	1.7	0.062	0.17	2.3
46	0.33	0.57	1.2	2.4	0.081	0.12	1.8
55	0.33	0.69	1.5	2.9	0.11	0.098	1.7
80	0.29	0.90	2.5	3.8	0.14	0.086	1.4
111	0.35	1.4	3.4	6.2	0.26	0.11	1.6
			T_1 (22.6 MHz), s				
10	0.13	0.20	0.45	1.1	0.025	a	0.69
22	0.11	0.25	0.52	1.3	0.030	0.034	0.64
49	0.094	0.33	0.89	1.7	0.048	0.031	0.54
83	0.10	0.45	1.1	2.1	0.079	0.035	0.53
105	0.15	0.67	1.4	2.8	0.13	0.043	0.70

a Not evaluated.

Table II. Nuclear Overhauser Enhancement Factors for Poly(n-butyl methacrylate) in 50% (w/w) Solution in Toluene-d_8

Temp. °C	NOEF (67.9MHz)a						
	C-1	C-2	C-3c	C-4	CCH_3	CCH_2	CCR_4
-5	2.0	1.9		2.0	b	b	0.51
14	1.4	1.7		1.8	0.71	0.67	0.47
42	1.1	1.6		1.5	1.2	0.43	0.66
52	0.80	1.4		1.5	1.2	0.44	0.50
80	0.54	1.3		1.3	1.3	0.43	0.56
101	0.42	0.98		1.3	1.4	0.36	0.59
			NOEF (22.6 MHz)a				
11	0.47	1.5		1.6	0.99	b	0.89
21	0.54	1.7		1.7	b	0.30	0.88
50	0.69	1.2		1.6	1.8	0.40	0.68
81	0.9	1.1		1.4	1.8	0.66	1.3
104	1.1	1.3		1.4	1.4	0.59	1.1

a NOE=NOEF +1; NOEF$_{max}$=1.99; extimated maximum error ±10%.
b Not evaluated. c Not evaluated due to overlap with solvent resonance.

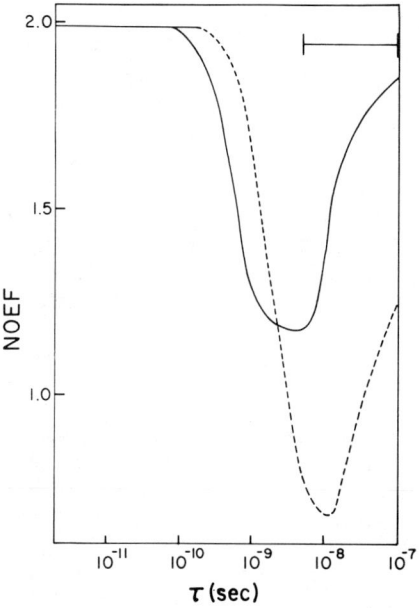

Figure 1. NOEFs for C-1 carbon undergoing internal rotation as a function of the isotropic correlation time for backbone motion at 22.6 (- - -) and 67.9 MHz (———). $D_i = 1 \times 10^{10}$ s^{-1}. The single correlation time model is used (8).

Poly(n-hexyl methacrylate). Although there is significantly less motion in PHMA than in PBMA, T_1 and NOEF trends are similar for the two polymers. As can be seen in Table III, the T_1's of C-1 in PHMA at 67 MHz are again ~0.30 sec over the 100° temperature range studied. The T_1 minima of C-1 and C-CH$_2$ occur at a temperature too high to observe in toluene at high field, but at low field the relatively shallow minimum again gives a minimum T_1 which is approximately twice that predicted from a single correlation time model. Also it is interesting to note that the T_1 of C-2 at 3° and 67 MHz ia again shorter than C-1, similar to results for the PBMA system. Comparison of the T_1's at the two fields reveals that there is still a significant field dependence even as far as C-6 of the alkyl chain. NOEF trends for PHMA are similar to those observed for PBMA.

Poly(n-butyl acrylate). A study of the relaxation properties of PBA was initiated for several reasons. There are two backbone carbons with directly bonded protons; thus the effect of the side chain on backbone motion might be determined. Also, the CH carbon should more directly reflect the distribution of correlation times necessary to begin analysis of alkyl sidechain motion. Finally, the lack of the additional chain-CH$_3$ groups significantly loosens motional constraints in PBA. The effect of this on the overall dynamics of PBA was of interest.

Poly(n-butyl acrylate) T_1's and NOEF's are given in Tables IV and V respectively. Although there is considerably more motion in this polymer compared with PHMA and PBMA, the same qualitative relaxation phenomena are still observed (i.e. low NOEF's and field dependent T_1's). The T_1 behavior of PBA at a given temperature is generally similar to that of PBMA and PHMA at 20-30° higher temperature. The T_1's of the backbone carbons are relatively insensitive to temperature, although PBA has apparently already passed the T_1 minimum at 40° even at 22.6 MHz. The T_1 properties of the butyl group are similar to those of the side chains in the other two polymers, although the field dependence is not as large (T_1 67MHz/T_1 22MHz≈1.3 compared with ≈2 for PBMA and PHMA). The NOEF's of the side chain carbons of PBA, and especially C-4, are surprisingly low. This may be a result of weighted contributions from rapid internal alkyl group rotations and overall motion but this degree of NOEF reduction has not been observed in PBMA or PHMA.

1,2-Decanediol, 1,2-Hexadecanediol, and 1,2,3-Decanetriol. Extensive inter- and intra-molecular hydrogen bonding in these compounds results in motional restrictions analogous to those present in polymer chains. Unlike the polymers, these polyols can be readily prepared with a variety of alkyl chain lengths, allowing us to study the extent of propagation of motional complexities through the chain.

The T_1's of DD are shown in Table VI. Except for the restricted end of the molecule at low temperature, all the T_1's

Table III. ^{13}C Spin-Lattice Relaxation Times of Poly(n-hexyl methacrylate) in 50% (w/w) Solution in Toluene-d_8

Temp. °C	C-1	C-2	C-3	C-4	C-5	C-6	CCH$_3$	CCH$_2$	CCR$_4$
					T_1 (67.9MHz), s				
34	0.30	0.30	0.55	1.1	1.7	2.6	0.068	0.13	1.8
49	0.31	0.47	0.85	1.6	2.4	3.7	0.092	0.11	1.6
63	0.31	0.54	1.1	2.1	3.2	4.6	0.10	0.097	1.6
83	0.30	0.65	1.4	2.3	3.7	5.3	0.12	0.090	1.5
100	0.31	0.81	1.8	2.9	5.1	7.1	0.22	0.093	1.4
					T_1 (22.6 MHz), s				
43	0.11	0.24	0.47	0.90	1.5	2.3	0.037	0.035	0.54
64	0.10	0.28	0.63	1.1	2.1	3.3	0.057	0.029	0.40
89	0.10	0.38	0.95	1.7	3.0	4.8	0.086	0.030	0.47
102	0.13	0.52	1.0	1.8	3.3	5.1	0.14	0.042	0.58

Table IV. ^{13}C Spin-Lattice Relaxation Times of Poly(n-butyl acrylate in 50% (w/w) Solution in Toluene[a].

Carbon		1	2	3	4	CH_2	CH
T(°C)	ν(MHz)						
40	22.6	0.98	2.3	3.4	11	0.09	0.25
55		0.95	2.6	4.0	13	0.22	0.56
70		1.4	3.0	5.0	16	0.14	0.52
40	67.9	1.4	2.8	4.0	13	0.22	0.56
55		1.6	3.2	4.4	14	0.23	0.67
70		2.0	4.0	5.5	13	0.26	0.75

[a] T_1's in s, ±5-10%

Table V. ^{13}C NOEF's of Poly(n-butyl acrylate) in 50% (w/w) Solution in Toluene[a].

Carbon		1	2	3	4	CH_2	CH
T(°C)	ν(MHz)						
40	22.6	1.0	1.0	0.7	0.23	1.4	1.1
55		1.8	1.6	1.9	0.23	1.1	1.6
70		1.2	1.5	1.4	0.28	1.1	1.7
40	67.9	1.8	1.7	1.8	0.50	0.95	1.3
55		1.9	1.9	1.9	0.61	1.2	1.5
70		1.7	1.6	1.7	0.50	1.3	1.4

[a] ±10-20%

Table VI. Spin Lattice Relaxation Times for Neat 1,2-Decanediol.

Carbon		1	2	3	4	5	6	7	8	9	10
T(°C)	ν(MHz)										
50	22.6[a]	0.26	0.40	0.24	0.31	0.41	0.50	0.63	0.96	1.6	4.1
60		0.35	0.49	0.27	0.35	0.45	0.55	0.72	1.2	2.0	4.6
75		0.55	0.77	0.42	0.52	0.58	0.72	1.1	1.8	2.6	5.1
92		1.1	1.3	0.85	1.1	1.2	1.2	1.6	3.1	4.9	7.1
112		2.0	2.9	1.7	1.9	2.3	2.6	3.1	5.2	7.3	10.3
50	67.9[b]	0.48	0.73	0.47	0.58	0.66	0.89	1.2	1.6	2.5	4.7
58		0.56	0.86	0.53	0.65	0.73	0.98	1.3	1.8	2.8	5.3
75		0.80	1.2	0.78	0.92	1.1	1.4	1.8	2.5	3.8	6.6
90		1.2	1.7	1.0	1.2	1.4	1.7	2.3	3.2	4.8	8.1
110		1.7	2.5	1.5	1.7	1.9	2.4	3.1	4.3	6.3	9.0

[a] T_1's at 22.6 MHz ±10%

[b] T_1's at 67.9 MHz ±5%

are characteristic of those in the extreme spectral narrowing region. Furthermore, as the temperature is raised, T_1's increase. Although there is significantly more motion for the diol than there is in the case of the polymers, a dipolar T_1 field dependence is still evident even to the end of the alkyl chain, which is equivalent to the sixth-eighth alkyl carbon in poly(alkyl methacrylates). The CH_2 carbon T_1 ratio C-9:C-3 is high, indicative of the high molecular association which occurs even at 95°. As with the polymers, NOEF's of DD (Table VII) are still not predicted by simple theroy, although they are larger and show temperature dependences opposite to that observed for the polymers at high field. Presumably, the field dependences observed for decanediol are a result of complex molecular association and/or micelle formation in these solutions. When the autocorrelation function becomes nonexponential, the effect is propagated along the pendant alkyl chain.

In order to determine the extent of propagation of this correlated motion, the T_1's of hexadecanediol were obtained (Table VIII). The results indicate that the motional characteristics responsible for field dependent relaxation are present to C-16 at 75°. This behavior disappears between 80° and 95°. Although it is desirable to extend these measurements to longer alkyl chains, even for HDD, resolution of C-5 through C_{n-3} is not possible, and only composite relaxation curves are obtained.

1,2,3-Decanetriol should show more extensive intermolecular hydrogen bonding than either DD or HDD. This is consistent with the significantly shorter T_1's observed for DT relative to DD. Although the T_1's are only marginally field dependent at best for the functionalized carbons (except at 50°), it is significant that T_1's vary by as much as a factor of two for the remainder of the carbons even as high as 75° (Table IX). The NOEF's of DT are all much lower (NOEFs ≈1.0 to 1.5) than expected from simple theory below 95°, where the T_1 field dependence vanishes and NOEF's approach 2.0.

<u>Polymer Backbone Motion</u>. Alternate descriptions of molecular motion utilize an effectively non-exponential autocorrelation function to describe polymer dynamics. One formalism is the use of a log-χ^2 distribution of correlation times in place of a single correlation time(14). Such a description may simulate the various time scales for overall and internal motions in polymers.

When a log-χ^2 distribution is invoked, ^{13}C T_1's and NOE's are affected as in Figures 2 and 3. In those figures, the parameter <u>p</u> describes the width of the distribution, with small values corresponding to a broad distribution, and $p = \infty$ corresponding to the single correlation time model.

A wide distribution (p≅8) causes the T_1 minimum to become shallow, and increases the value of the minimum T_1 value which will be observed. The NOEF (Figure 3) is reduced at short correlation times, and larger than expected at long correlation

Table VII. ^{13}C NOEF's^a for Neat 1,2-Decanediol.

Carbon		1	2	3	4	5	6	7	8	9	10
T(°C)	ν(MHz)										
50	22.6	2.0	1.8	1.3	1.3	1.3	1.4	1.4	1.3	1.0	1.7
60		1.7	1.5	1.2	1.0	1.2	1.1	1.1	1.0	0.7	1.4
75		1.8	1.9	1.3	1.2	1.2	1.2	1.1	1.1	0.8	1.8
92		1.6	1.4	0.7	0.8	1.0	0.9	0.8	0.8	0.6	1.3
112		2.1	1.0	1.0	0.9	1.0	1.0	0.9	0.9	0.7	1.8
50	67.9	1.8	1.7	1.5	1.5	1.5	1.6	1.6	1.6	1.7	1.8
58		1.7	1.6	1.5	1.5	1.5	1.5	1.5	1.6	1.7	1.8
75		1.9	1.7	1.6	1.6	1.5	1.8	1.9	1.7	1.8	1.7
90		2.0	1.9	1.9	1.9	1.9	1.9	1.9	1.8	1.8	1.8
110		1.8	1.9	1.9	1.8	1.6	1.7	1.6	1.5	1.5	1.6

^{a} ±10–15%

Table VIII. ^{13}C Spin Lattice Relaxation Times (T_1) for Neat 1,2-Hexadecanediol.[a]

T(°C)	ν(MHz)	Carbon Number							
		1	2	3	4	5-13		14	15	16
75°	22.6[a]	0.18	0.16	0.23	0.26	0.66	0.42	0.65	1.3	2.3	5.6
	67.9[b]	0.50	0.62	0.35	0.51	0.85	→	1.5	2.2	3.6	6.5
80°	22.6[a]	0.30	0.29	0.51	0.48	—	1.1	—	2.1	2.8	6.0
	67.9[b]	0.66	0.76	0.60	0.65	0.9	→	1.5	2.5	4.1	7.5
95°	22.6[a]	0.76	0.74	1.3	1.1	—	1.7	—	3.2	6.4	8.5
	67.9[b]	0.66	0.74	1.2	1.0	1.1	→	1.7	3.4	7.0	8.1

[a] ±10-15% error
[b] ±10% error

Table IX. ^{13}C Spin Lattice Relaxation Times for Neat 1,2,3-Decanetriol^a.

Carbon		1	2	3	4	5	6	7	8	9	10
T	ν(Hz)										
50°	22.6	0.08	0.10	0.09	0.09	0.10	0.18	0.20	0.48	0.70	1.9
	67.9	0.11	0.17	0.19	0.10	0.21	0.34	0.43	0.71	1.1	2.7
60°	22.6	0.15	0.20	0.18	0.13	0.17	0.25	0.34	0.55	0.97	2.5
	67.9	0.15	0.22	0.22	0.19	0.26	0.40	0.52	0.84	1.5	3.2
75°	22.6	0.28	0.30	0.35	0.21	0.22	0.30	0.50	0.87	1.6	4.1
	67.9	0.26	0.36	0.38	0.25	0.36	0.53	0.69	1.1	1.8	4.2
95°	22.6	0.77	0.91	0.85	0.54	0.64	0.87	1.2	1.9	3.2	6.5
	67.9	0.65	0.81	0.75	0.52	0.66	0.86	1.2	1.7	3.0	6.0

^{a}±10%, T_1's are the average of those obtained for RR/SS and RS/SR diastereomers, which in all cases differed by less than 7%.

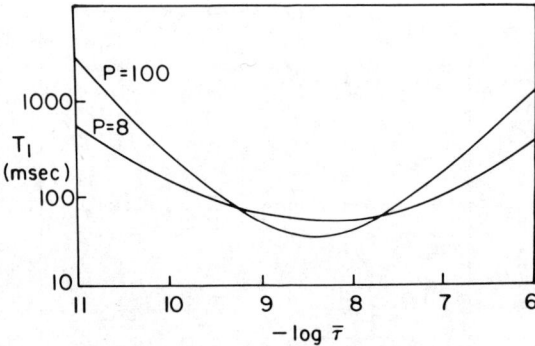

Journal of the American Chemical Society

Figure 2. Spin-lattice relaxation time, T_1, as a function of correlation time, τ, and distribution width, p, for the log-χ^2 distribution (22.6 MHz) (8)

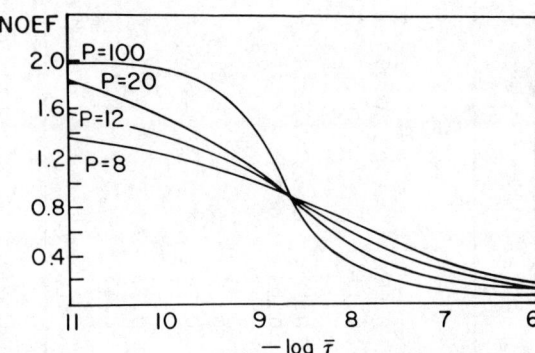

Journal of the American Chemical Society

Figure 3. NOEF as a function of correlation time and distribution width for the log-χ^2 distribution (22.6 MHz) (8)

times, when a broad distribution is described. From Figure 4, it can also be seen that use of a distribution of correlation times also introduces field dependent behavior for T_1's on the left side of the T_1 minimum. These trends are all observed for the polymer and polyol samples.

The observed and calculated T_1's and NOEF's of PBMA backbone carbons are illustrated in Table X. The agreement between the two sets of numbers is quite good, considering the approximations that are inherent in the model. Not only the field dependence, but also the temperature dependence of the relaxation parameters is predicted quite well.

Distributions of correlation times can also be applied to the non-polymeric systems. The predicted T_1's and NOEF's of the restricted end of DT, both with and without a distribution invoked, are shown in Table XI. In decanediol motion is not as restricted as in the polymers, leading to shorter τ's and narrower distributions. Analysis of these motionally restricted backbone carbons yields overall correlation times which can be used as a basis for further analysis of the pendant alkyl groups.

<u>Pendant Alkyl Chain Motion</u>. Although the relaxation properties of DT side chain carbons can be predicted qualitatively by using a distribution of correlation times, this type of motion is not physically realistic. A more realistic model for the segmental motion of these pendant groups involves superposition of separate C-C bond internal rotations and librations on overall molecular motion. This model works quite well for analyzing single frequency data. The relaxation times for C-4 through C-7 at 22.6 MHz are accurately predicted by internal C-C bond rotation with diffusion constants $D_i \approx 1 \times 10^9$ (i=1-4) superimposed on overall molecular motion with $\tau_c = 5 \times 10^{-10}$ sec (Figure 5). The relaxation of carbons, 8,9 and 10 are governed by a faster internal diffusion constant. These same motional parameters <u>do not predict the relaxation properties at 67.9 MHz</u> (Figure 6).

Better success is achieved by combining a distribution of correlation times with multiple internal rotations for the alkyl groups. Calculated relaxation data using this combination for C-4, 5, and 6 of DT at 50° using various distributions are presented in Table XII. The single correlation time model (p=∞) does not predict the T_1 field dependence and reduced NOEF's that are observed. An approximate fit of the calculated relaxation parameters to those observed for C-4, 5 and 6 at 50° can be obtained assuming overall molecular motion to occur with $\bar{\tau} = 5 \times 10^{-10}$ sec and $D_1 = D_2 = D_3 = 1 \times 10^{10} \text{sec}^{-1}$ (the diffusion coefficients for rotation about the C-3/C-4, C-4/C-5, and C-5/C-6 bonds respectively) with p=8. Similarly the relaxation of the same carbons at 60° can be approximated with motion described by $\bar{\tau} = 2 \times 10^{-10}$ sec, $D_1 = D_2 = D_3 = 1 \times 10^{10} \text{ sec}^{-1}$, and p=8.

There may be some concern since relatively constant D_i values satisfactorily predict the observed relaxation behavior for

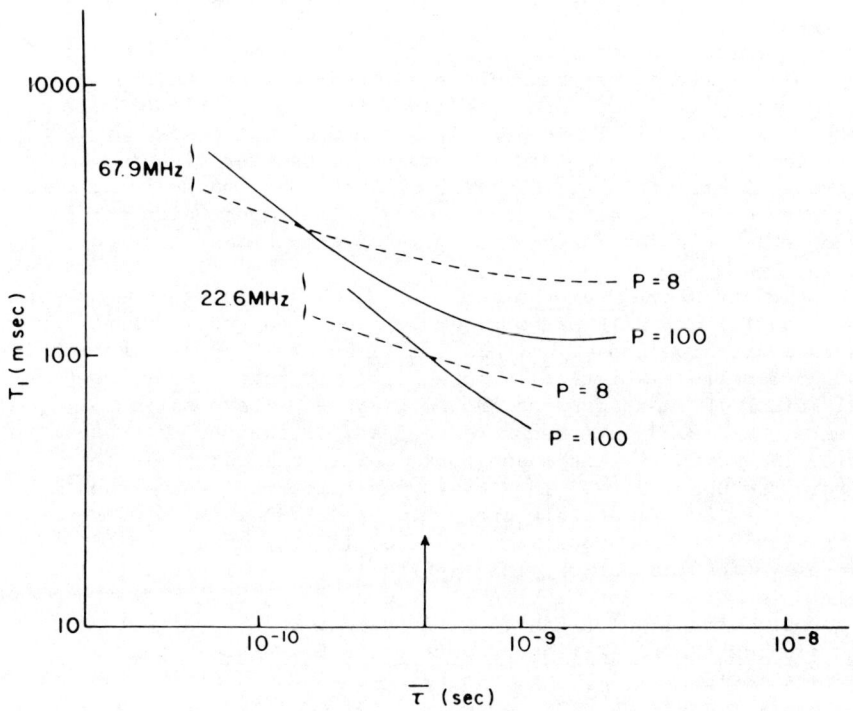

Journal of the American Chemical Society

Figure 4. Spin-lattice relaxation time as a function of correlation time for distribution widths of p = 8 (- - -) *and* p = 100 (———) *for 67.9 and 22.6 MHz (8)*

Table X. Observed and Calculated NT_1 and NOEF values for PBMA (50% in Toluene-d_8) as a Function of Temperature Assuming a Log -χ^2 Distribution.

Temp. °C	67.9 MHz				22.7 MHz				$\bar{\tau} \times 10^9$ s	Calcd width parameter
	NT_1 (s)		NOEF		NT_1 (s)		NOEF			
	Calcd	Obsd	Calcd	Obsd	Calcd	Obsd	Calcd	Obsd		
100	0.19	0.19	0.37	0.36	0.059	0.079	0.59	0.59	7.1	20
80	0.17	0.17	0.43	0.43	0.051	0.069	0.67	0.66	5.0	20
50	0.22	0.22	0.41	0.44	0.059	0.063	0.59	0.40	11	12
40	0.25	0.25	0.45	0.43	0.068	0.062	0.61		16	8

Table XI. Calculated Carbon-13 T_1's and NOEF's of 1,2,3-Decanetriol at 50° Using a Single τ or a Distribution of τ's.

Carbon	67.9 MHz		22.6 MHz			
	T_1, s	NOEF	T_1, s	NOEF	p	$\bar{\tau}^a$
1	0.11	1.2	0.08	1.6	b	
	0.11	1.8	0.09	2.0	∞	2.3^c
	0.11	1.4	0.08	1.7	50	2.3
2	0.17	0.7	0.10	1.4	b	
	0.17	1.5	0.14	2.0	∞	3.7^c
	0.17	1.0	0.09	1.4	30	4.5
3	0.19	0.6	0.09	1.4	b	
	0.19	1.5	0.14	2.0	∞	3.7^c
	0.19	1.0	0.09	1.4	30	4.5

$^a\bar{\tau}$ given in units of 10^{-10} sec.

^{b} Experimental values.

^{c} Equivalent to use of a single τ.

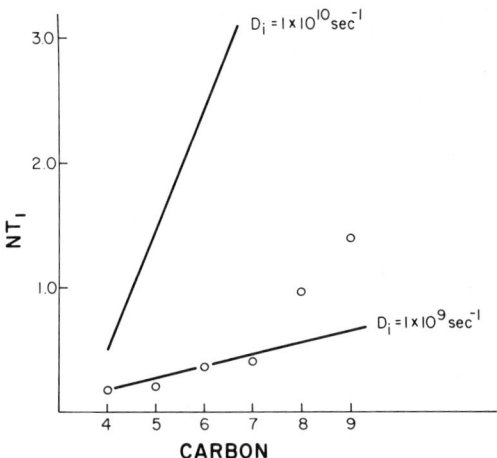

Figure 5. NT_1 as a function of carbon number at 22.6 MHz for the alkyl carbons of DT. Carbon-3 is assumed to be the effective center of mass: (——) calculated using the MIR Theory with a single correlation time; (○) experimental points for DT at 50° and 22.6 MHz; $\bar{\tau} = 5 \times 10^{-10}$ s.

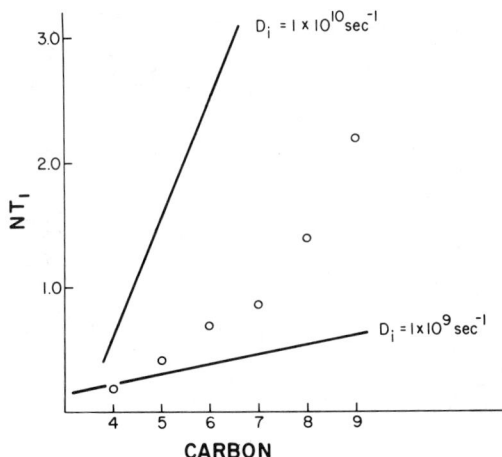

Figure 6. Same as Figure 5 except that data is for 67.9 MHz

Table XII. Calculated Carbon-13 T_1's and NOEF's Using the Theory of Multiple Internal Rotations Both With and Without a Distribution of Correlation Times for Overall Motion.

$\bar{\tau}^a$	$D_i(s^{-1})^b$	p	Carbon - 4				Carbon - 5				Carbon - 6			
			67.9 MHz		22.6 MHz		67.6 MHz		22.6 MHz		67.6 MHz		22.6 MHz	
			T_1	NOEF	T_1	NOEF	T_1	NOEF	T_1	NOEF	T_1	NOEF	T_1	NOEF
5×10^{-10}	10^9	8	0.16	0.9	0.07	1.0	0.16	0.9	0.07	1.1	0.17	1.0	0.07	1.1
		20	0.13	1.0	0.06	1.2	0.14	1.2	0.07	1.3	0.15	1.2	0.09	1.4
		50	0.12	1.2	0.07	1.5	0.15	1.3	0.09	1.6	0.16	1.5	0.12	1.7
		∞	0.11	1.7	0.10	1.9	0.15	1.8	0.13	2.0	0.19	1.9	0.18	2.0
	10^{10}	8^c	0.24	1.1	0.12	1.2	0.29	1.2	0.16	1.3	0.35	1.2	0.19	1.3
		20	0.25	1.3	0.15	1.5	0.37	1.5	0.26	1.6	0.50	1.6	0.39	1.7
		50	0.28	1.5	0.18	1.5	0.50	1.8	0.44	1.8	0.80	1.9	0.80	1.9
		∞	0.31	1.7	0.26	1.9	0.75	1.9	0.70	2.0	1.25	2.0	1.25	2.0

D_1[b]													
2×10^{-10}	10^9	8	0.16	0.9	0.07	1.0	0.18	0.9	0.07	1.0	0.19	1.0	0.08	1.1
		20	0.13	1.0	0.06	1.2	0.15	1.2	0.09	1.3	0.17	1.2	0.09	1.4
		50	0.16	1.2	0.07	1.5	0.16	1.4	0.11	1.7	0.19	1.5	0.14	1.7
		∞	0.18	1.9	0.16	2.0	0.22	1.9	0.20	2.0	0.26	2.0	0.25	2.0
	10^{10}	8[d]	0.25	1.1	0.12	1.2	0.29	1.2	0.16	1.3	0.35	1.2	0.20	1.3
		20	0.27	1.3	0.14	1.4	0.39	1.5	0.27	1.7	0.50	1.6	0.29	1.7
		50	0.32	1.6	0.23	1.6	0.55	1.8	0.49	1.9	0.8	1.9	0.8	2.0
		∞	0.43	1.9	0.43	2.0	0.54	2.0	0.54	2.0	1.4	2.0	1.4	2.0

[a] Estimated using simple theory modified for log-χ^2 distribution of τ_c's for carbons 1, 2, and 3.

[b] D_i is rotational diffusion constant for rotation about a c–c bond, D_i is the constant for rotation about C_3–C_4, D_2 is the constant for rotation about C_4–C_5, etc. D_i is assumed to be constant here, for the first four carbons (i.e., $D_1=D_2=D_3=D_4=1 \times 10^{10}$ or $1 \times 10^9 \text{sec}^{-1}$).

[c] Predicts relaxation behavior of 1,2,3-decanetriol at 50°.

[d] Predicts relaxation behavior of 1,2,3-decanetriol at 60°.

several chain carbons. At first glance it appears that methyl, ethyl, propyl, etc. group rotations are equivalent, a situation which is not physically realistic. However, for the internal portion of the chain, there should be little difference in the ability of the entire length of a n-pentyl or a n-hexyl chain to rotate. Furthermore, carbon-carbon bond rotations near the site of restriction do not necessarily involve movement of the entire remaining portion of the chain; they could involve simultaneous motion about several carbon-carbon bonds. This type of motion is probably characteristic of the internal carbons (e.g. C-4 through C-7 of 1,2,3-decanetriol). Carbons 8 and 9 of 1,2,3-decanetriol can involve rotation of propyl and ethyl groups; their larger diffusion constants probably reflect contributions from this type of motion. These are only approximations; better fits can be obtained by adjustment of D_i and p parameters but of course such fitting procedure may lose significance. The fit of C-4 relaxation data is not as good as those for the other carbons. This may result from conformationally restricted motion for C-4, with gauche interactions for C-5 with the vicinal -OH, and -CH(OH)-CH_2OH group limiting rotation around the C-3/C-4 bond.

Similar extensive calculation have also been performed with the data from PBMA and PHMA(8). Selected data for C-1 of the poly(n-alkyl acrylates) are shown in Table XIII. Rather than trying to closely fit each of the data sets with variable parameterization of $\bar{\tau}$, p, and D_i, values were chosen to encompass the relaxation characteristics of the entire temperature range used in the study. The same insensitivity of T_1 to temperature variation observed in the experimental data is also seen in the calculated data when $\bar{\tau}$ is changed by two orders of magnitude (Table XIII). Even with a relatively narrow distribution (p=50) a large T_1 field dependence and low NOEF is observed.

Sample calculations for C-2 are shown in Table XIV. The experimental conditions for which the calculated and experimental relaxation data agree are shown in the right columns. For similar temperatures, the motion of PBMA is freer than PHMA. The diffusion coefficients for C-C bond rotation increase only slightly if at all, as the separation from the anchored end of the chain becomes larger. This trend is observed all the way to C-5 of PHMA; C-6 is a methyl carbon with higher symmetry, expected to rotate more rapidly. Although the NOEF behavior of C-2 is not accurately predicted in Table XIV, the observed NOEF temperature dependence is probably a result of a mix of various types of motion analogous to that observed for C-1 (Figure 1).

In view of the large number of parameters which are variable (p $\bar{\tau}$, and D_i's) the necessity of having at least the four parameters (i.e., T_1's and NOEF's at two fields) is imperative, since in some instances two or more combinations of motional parameters can closely predict T_1's. Data obtained at three widely spaced fields would be better. Further, a knowledge of the temperature dependence of these relaxation parameters is necessary.

Table XIII. Calculated Methylene Spin-Lattice Relaxation Times and Nuclear Overhauser Enhancement Factors for the C-1 Carbon

Disbribution width[a]	67.9 MHz		22.6 MHz	
	T_1, s	NOEF	T_1, s	NOEF
$P=8$ ($\bar{\tau}=1.3 \times 10^{-7}$ s)				
$D_1=1.2$	0.24	1.1	0.12	1.1
$D_1=3.0$	0.34	1.1	0.16	1.1
$P=20$ ($\bar{\tau}=5.0 \times 10^{-9}$ s)				
$D_1=1.2$	0.26	1.3	0.14	1.3
$D_1=3.0$	0.40	1.3	0.21	1.2
$P=50$ ($\bar{\tau}=5.0 \times 10^{-9}$ s)				
$D_1=1.2$	0.29	1.3	0.15	1.2
$D_1=3.0$	0.46	1.2	0.21	1.0

[a] Log -χ^2 distribution assumed: all D_1's are given in units of 10^{10} s^{-1}.

[b] For PBMA and PHMA. Listed diffusion coeeficients reproduce observed data for the stated experimental temperatures.

Table XIV. Calculated Methylene Spin-Lattice Relaxation Times and Nuclear Overhauser Enhancement Factors for the C-2 Carbon

Distribution width[a]	67.9 MHz T_1, s	67.9 MHz NOEF	22.6 MHz T_1, s	22.6 MHz NOEF	Exptl temp.°C[b] PBMA	Exptl temp.°C[b] PHMA
$P = 8$						
($\bar{\tau} = 1.3 \times 10^{-7}$ s)						
$D_1 = 1.2, D_2 = 3.0$	0.36	1.2	0.20	1.3		
$D_1 = 1.2, D_2 = 5.0$	0.41	1.2	0.22	1.3	21	34
$D_1 = 1.2, D_2 = 10.0$	0.48	1.3	0.26	1.3		
$P = 20$						
($\bar{\tau} = 5.0 \times 10^{-9}$ s)						
$D_1 = 1.2, D_2 = 3.0$	0.52	1.6	0.36	1.6	46	63
$D_1 = 1.2, D_2 = 5.0$	0.62	1.6	0.44	1.6	55	83
$D_1 = 1.2, D_2 = 10.0$	0.75	1.6	0.53	1.6		100
$P = 50$						
($\bar{\tau} = 5.0 \times 10^{-9}$ s)						
$D_1 = 1.2, D_2 = 1.0$	0.51	1.8	0.40	1.7		
$D_1 = 1.2, D_2 = 3.0$	0.78	1.8	0.60	1.6		
$D_1 = 1.2, D_2 = 5.0$	0.92	1.7	0.69	1.6	98	
$D_1 = 1.2, D_2 = 10.0$	1.13	1.7	0.81	1.5		

[a] Log -χ^2 distribution assumed; all D_1's are given in units of 10^{10} s^{-1}.
[b] Listed diffusion coefficients reproduce observed data for the stated experimental temperatures.

While incorporation of a distribution is a good solution to the problem of molecular relaxation in polymeric systems, it is not unique. Howarth(17a,b) and Bull(17c) have invoked librational motions to explain nmr relaxation phenomena in several peptides and proteins; however, the initial work utilized many variables. Recently (17b), Howarth has used this same theory to analyze our earlier polymer (8) and preliminary decanediol (10) data. By varying two correlation times, one for overall molecular motion and one for local segmental motion and using a librational angle of $\theta \approx 20°$ he was quite successful in analysing the data for 1-decanol, DD, PHMA, PBMA, and random coil poly-γ-benzy-L-glutamate. (It should be noted however that the Howarth treatment is not entirely satisfactory since even slight adjustment of the variables can have profound consequences.)

Conclusion

It is evident that simple theories of molecular motion are not adequate to explain experimental nmr relaxation parameters in certain polymer systems as well as in some highly associated small molecules. As field dependent nmr relaxation studies become more widespread, the observation of these relaxation characteristics will undoubtedly be found more general than is currently thought.

With the large number of motional theories being touted the need for multi-frequency relaxation studies becomes critical. At one frequency most theories can satisfactorily predict the T_1 behavior because of the many adjustable parameters. By initiating multifield and multitemperature T_1 and NOEF studies, more subtle features of molecular motion can be probed. Although the motional model used by us is adequate, it may not be the best model. Indeed, Howarth has had better results with our preliminary data using internal librational motion. This enforces the need for measuring as many relaxation parameters as possible, under as many different conditions as possible.

Acknowledgements

The authors are grateful to the National Science Foundation for financial support of this work (Grant CHE-77-26473 to GCL), and to the National Institutes of Health Postdoctoral Fellowship Program (Grant 1-F32-CA06370-01 to PLR).

Literature Cited

1. See references cited in: Wright, D.A.; Axelson, D.E.; Levy, G.C., "Topics in Carbon-13 NMR Spectroscopy" G.C. Levy, Ed., Wiley-Interscience, New York, Chapter 2, 1979.

2. (a) Woessner, D.E. J. Chem. Phys., 1962, 36, 1; (b) Woessner, D.E. J. Chem. Phys., 1962, 37, 647.

3. Wilbur, D.J.; Norton, R.S.; Clouse, A.O.; Addlemar, R.; Allerhand, A. J. Am. Chem. Soc., 1976, 98, 8250.

4. Llinas, M.; Meiey, W.; Wüthrich, K. Biochem. Biophys. Acta, 1977, 492, 1.

5. Visscher, R.B.; Gurd, F.R.N. J. Biol. Chem., 1975, 250, 2238.

6. Nelson, D.J.; Opella, S.J.; Jardetzky, O. Biochem., 1976, 15, 5552.

7. Wüthrich, K.; Baumann, R. Org. Magnetic Resonance, 1976, 8, 532.

8. Levy, G.C.; Axelson, D.E.; Schwartz, R.; Hochmann, J. J. Am. Chem. Soc., 1978, 100, 410.

9. Heatley, F.; Begum, A. Polymer, 1976, 17, 399.

10. Levy, G.C.; Cordes, M.P.; Lewis, J.S.; Axelson, D.E. J. Am. Chem. Soc., 1977, 99, 5492.

11. Somorjai, R.L.; Deslauriers, R. J. Am. Chem. Soc., 1976, 98, 6760.

12. Lynch, L.J.; Marsden, K.H.; George, E.P. J. Chem. Phys., 1969, 51, 5673.

13. Tsutsumi, A; Chachaty, C. Macromolecules, 1979, 12, 429.

14. Schaefer, J. Ibid., 1973, 6, 882.

15. London, R.E.; Avitabile, J. J. Am. Chem. Soc., 1978, 100, 7159.

16. Wittebort, R.J.; Szabo, A. J. Chem. Phys., 1978, 69, 1722.

17. (a) Howarth, O.W., J.C.S., Faraday Trans II, 1978, 74, 1031.
    (b) Howarth, O.W., J.C.S., Faraday Trans II, 1979, 75, 863.
    (c) Bull, T.; Norne, J.E.; Reimarrson, R.; Lindman, B.
        J. Am. Chem. Soc., 1978, 100, 4643.

18. Dill, K.; Allerhand, A. J. Am. Chem. Soc., 1979, 101, 4377.

19. Cordes, M.P., M.S. Thesis, Chemistry Department, Florida State University, Tallahassee, Florida, 1978.

20. Rinaldi, P.L.; Levy, G.C., J. Org. Chem., manuscript submitted.

21. Canet, D.; Levy, G.C.; Peat, I.R. J. Magnetic Resonance, 1975, 18, 199.

22. Kowalewski, J.; Levy, G.C.; Johnson, L.F.; Palmer, L. J. Magnetic Resonance, 1977, 26, 533.

23. Levy, G.C.; Peat, I.R.; Rosanske, R.C. Ibid, 1975, 18, 205.

24. Kuhlman, K.F.; Grant, D.M.; Harris, R.K. J. Chem. Phys., 1970, 53, 3439.

RECEIVED June 3, 1980.

Use of Pulsed NMR to Study Composite Polymeric Systems

D. C. DOUGLASS

Bell Laboratories, Murray Hill, NJ 07974

The characterization of solid polymeric material often includes the need to characterize the variety of molecular motions present as well as the molecular and morphological structure. NMR relaxation measurements have a long history of application to molecular motion studies of polymers where NMR data often complements mechanical and dielectric measurements with a more complete identification of the mobile, or immobile, entities.

Since many useful commercial polymers are composite in the sense of being blends, segregated block copolymers, filled and plasticized material it becomes especially interesting to be able to identify which component is moving, its spatial extent and possibly the composition of a region.

The following NMR experiments relating to these questions were carried out in direct collaboration with Prof. Vincent J. McBrierty, Dublin University, Ireland. The great help and influence of several colleagues at the Bell Laboratories who have a notably active interest in composite materials,[1-3] T. K. Kwei, Harvey Bair and Henry Wang in particular, are also gratefully acknowledged.

While it is clear that molecular structure, molecular motion and morphology are not entirely separable features of a material, it is the purpose here to emphasize that NMR relaxation measurements can, in circumstances where spin diffusion or cross relaxation becomes partially rate controlling, be used to glean some pertinent information that may be assigned to the small scale morphology of polymers. To this claim one must add the perhaps obvious caveat that conclusions regarding spatial dimensions are qualitative in nature with presently available experiments. In contrast to solid state high resolution experiments, it is relatively easy to obtain reasonably accurate T_1, $T_{1\rho}$ and other data, data that often indicate the presence of spin diffusion. However, it is usually difficult to achieve better than a very rudimentary model of systems with complex morphology. With this in mind,

we turn to a brief outline of an appropriately qualitative
line of thought that underties both the visualization of
experiments and the interpretation of data.
   The concept of spin temperature is well established and
exploited in solid state NMR.[4,5] The notion is based upon two
generally valid features of an immobile spin system.
(1) The spins are very weakly coupled to their molecular
framework compared to the strength of their coupling to the
applied static magnetic field and to applied resonant radiation.
Indeed, resonant radiation can violently and coherently drive
the spin motions and provides the basis for all pulsed NMR
experiments. (2) In dense spin systems the strength of
interaction between neighboring spins is also large compared to
coupling to the host molecule or molecular lattice. These two
properties lead to two long-recognized consequences of interest
in this context, and schematically shown in figure 1; (1) the
system of spins may be assigned a temperature different from
that of the host lattice and relaxation may be thought of as
temperature equilibration of the lattice and a spin system
that has been cleverly heated or cooled by application of
resonant radiation and (2) dipolar coupling between neighboring
spins leads to an exchange of energy between neighbors and
hence to movement of energy within the spin system. If a
temperature gradient exists within the spin-system, energy
can move from the hot to the cold regions and this process is
referred to as spin-diffusion, often somewhat of a misnomer
because the energy, not a spin, moves, A somewhat quaint but
graphically appealing visualization of relaxation and
spin-diffusion is obtained by exploiting the analogy between
these processes and the flow of water amongst a system of
interconnected reservoirs, as shown in the schematic diagram
of figure 1.
   It is this spin diffusion phenomenon that is used to
provide an estimate of spatial dimensions in motionally
inhomogeneous systems. The idea simply being that observation
of the time required for equilibration of two regions in a
spin system gives a measure of the dimension involved in the
transport process, when the diffusion coefficient is known.
The initial temperature gradients may be established in two
general ways; (1) if some part of the sample relaxes energy
to the lattice rapidly it will cool and act as a sink for the
remainder of the spin system and (2) it is sometimes possible
to selectively heat a part of the spin system, as in Assinks'
use of the Goldman-Shen experiment.[6,7]
   The above qualitative picture will be applied to the
analysis of a model system, normal alkanes, as a test and then
we will proceed to consideration of blends of PSAN and PMMA,
a study of plasticized PVC, an example of cross relaxation in
PVF_2 - PMMA blends and finally a brief examination of blended
fibers of PE and PP.

The long-chain normal alkanes have a crystalline form in which the terminal methyl groups lie in a plane nearly normal to the chain axis.[8] At low temperatures the methyl group reorientation provides a rapidly relaxing site and it acts as a sink for the remainder of the proton spin system.[9] Figure 2 shows $T_{1\rho}$ data for a series of alkanes as a function of temperature. If spin diffusion were entirely rate controlling, the minimum would rise as the square of the number of methylene carbons, N; if relaxation to the lattice were the bottle neck, the minimum should rise linearly with N. The data exhibits a rise of the minimum with the 1.6 or 1.7 power of N, a typically intermediate case.

Figure 3 shows T_1 and $T_{1\rho}$ data on blends of PSAN copolymer with PMMA that results from a situation somewhat analogous to that of the alkanes.[10,11] In this case, the concentration of the methyl groups is altered by changing the concentration of PMMA in the blend (melt blended). At -100°C the methyl motion gives a pronounced $T_{1\rho}$ minimum. The corresponding T_1 minimum is at 0°C. If mixing is complete on a molecular scale one would expect a uniform spin temperature to prevail throughout the sample. Under this condition the observed rate of relaxation, k, is given by the relation.

$$k = k_1^\circ \frac{Nm}{N_T} + k_2^\circ \frac{N\phi}{N_T}$$

where k_1° and k_2° are the intrinsic relaxation rates of α-methyl and phenyl protons, and Nm/N_T and $N\phi/N_T$ are the fractions of α-methyl and phenyl protons. Assuming no excess volume of mixing, as has been shown to be valid for this system,[10] one should expect the relation

$$(7.1 + .88W_1)k = W_1 [3 k_1^\circ - 3.5 k_2^\circ] + 3.5 k_2^\circ$$

between the relaxation rate at the minima and the weight fraction PMMA, W_1, to be obeyed when mixing is complete and the system is not too dilute in methyl groups. Figure 4 gives T_1 versus composition. Within experimental error there is a linear dependence indicating that on the time scale of T_1 the transport is not rate controlling and the sample appears to be homogeneous. Figure 5 exhibits an analogous plot of $T_{1\rho}$ data which is clearly concave indicating that, on the shorter $T_{1\rho}$ time scale, spin diffusion is partially rate controlling. If the spin coefficient, D, is taken to be approximately 2×10^{-12} cm^2/sec., the mean square distance that spin energy may diffuse in a time like T_1 is given by

$$\sqrt{r^2} = \sqrt{T_1} \text{ Å,}$$

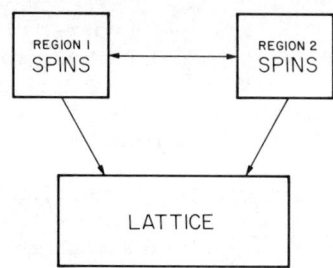

Figure 1. Schematic diagram of spin–spin and spin-lattice relaxation

Journal of Chemical Physics

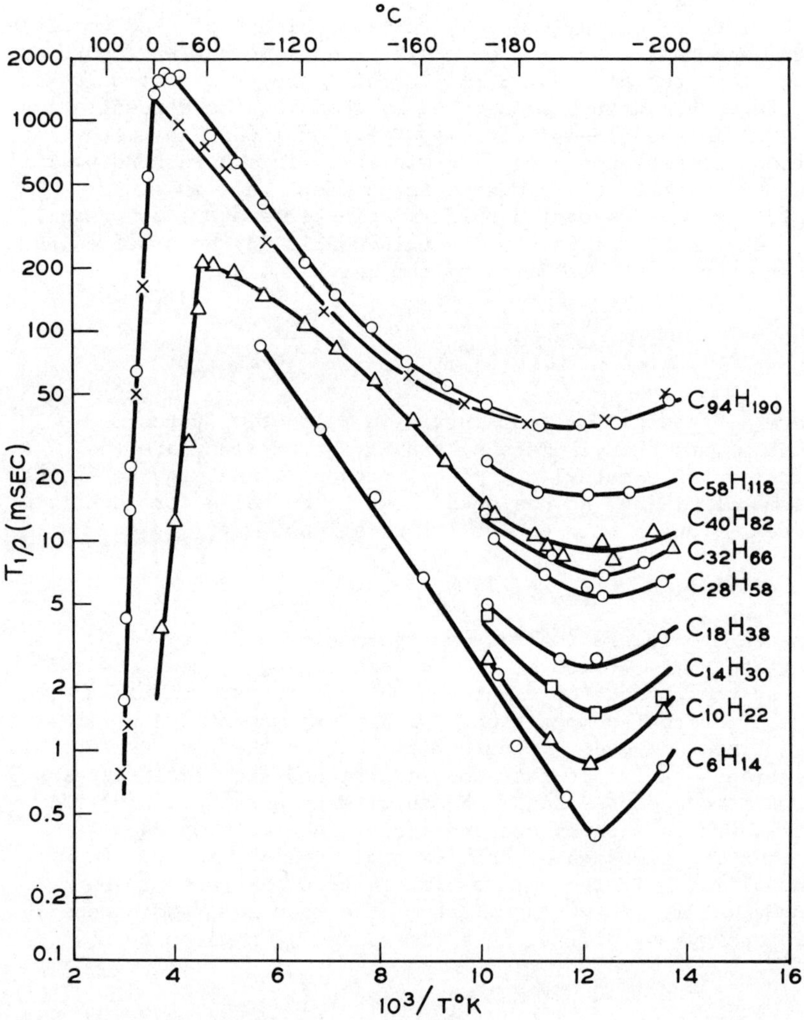

Figure 2. Rotating-frame relaxation data for normal alkanes vs. temperature (9)

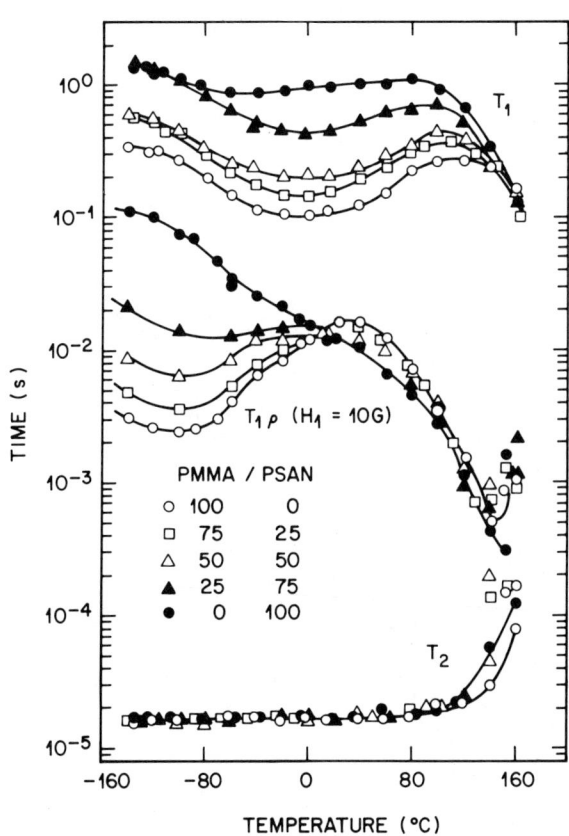

Macromolecules

Figure 3. T_1, T_2, and $T_{1\rho}$ data for PSAN/PMMA blends as a function of temperature (11)

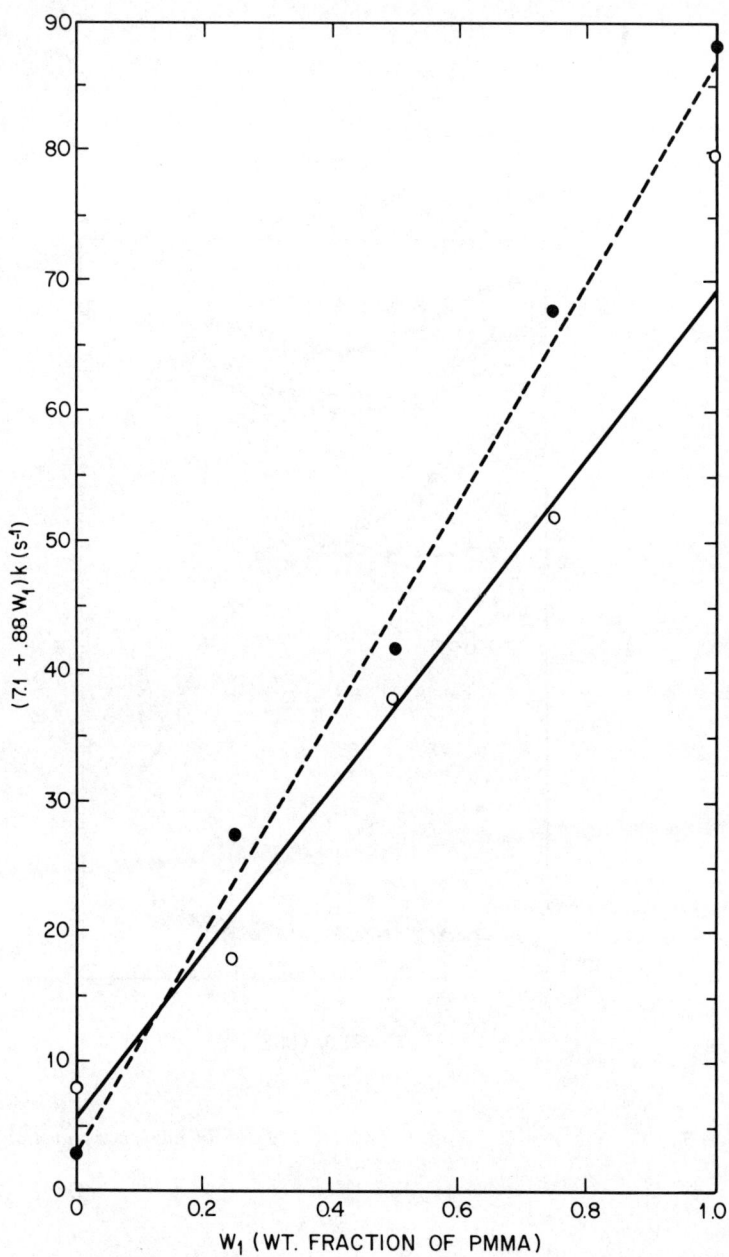

Macromolecules

Figure 4. Plot of $(7.1 + .88 w_1) k$ vs. w_1, where $w_1 =$ weight fraction of PMMA and $k = T_1^{-1}$. Note that the verticle axis of the corresponding figure of Ref. 11 is labeled incorrectly (11). T_1 data: (○) 40 MHz; (●) 30 MHz.

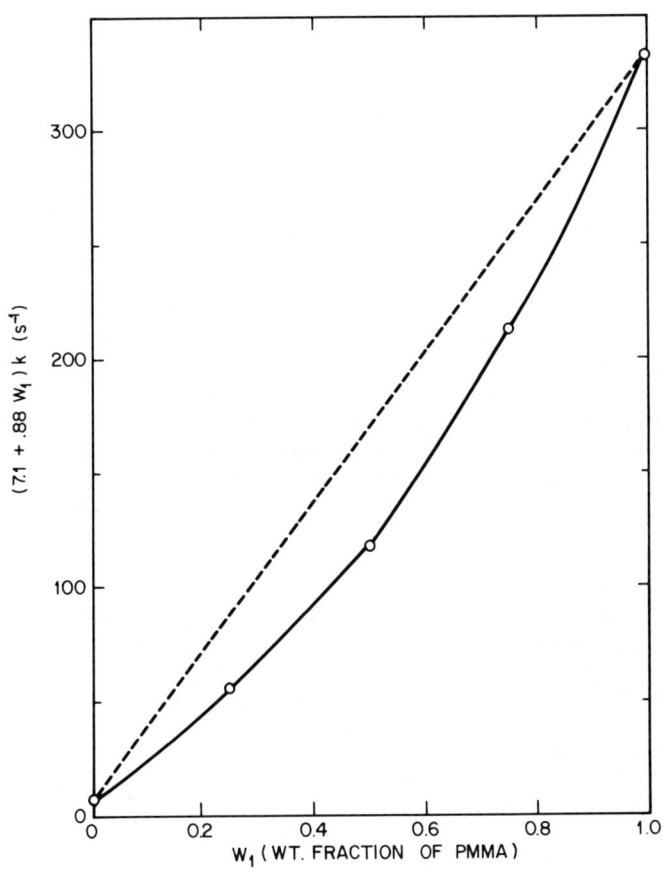

Macromolecules

Figure 5. Plot of $(7.1 + .88 \, w_1) \, k$ vs. w_1, where $w_1 =$ weight fraction of PMMA and $k = T_{1\rho}^{-1}$. Note that the verticle axis of the corresponding figure of Ref. 11 is labeled incorrectly (11). $T_{1\rho}$ data: $H_1 = 10G$; $k = 1/T_{1\rho}$.

or of the order of 150 Å. For the $T_{1\rho}$ time scale this distance is roughly 15 Å. Therefore, the blend appears to have some inhomogeneous structure of a size between 15 and 150 Å.

The results of an attempt to obtain alternative evidence from the shape of the magnetization decay in the $T_{1\rho}$ experiments is shown in figure 6. If the material is indeed inhomogeneous one would expect this decay to be a composite sum of at least two exponentials. For an RF field strength, H_1, of 10G no evidence of composite character is evident in the PSAN rich 75/25 sample at -100°C. However, if H_1 is decreased the intrinsic $T_{1\rho}$ of the methyl groups decreases with consequent relative increase in the role of spin diffusion. The decay plot for H_1 = 2G does exhibit curvature. It has been suggested by Dr. Garroway[12] that a composite decay may also develop as a result of other mechanisms if H_1 is sufficiently small. Therefore, this supporting evidence needs further experimental attention.

When H_1 is much larger than the local dipolar field, the shape of the $T_{1\rho}$ decay depends upon the small scale morphology of the inhomogeneities. Figure 7 shows the calculated composite decay resulting from a simple spherical domain having an exterior sink. Since composite character in the $T_{1\rho}$ decays depends upon geometric shape as well as scale, detailed analysis of these decay curves is challenging. But, in summary, while the absence of composite character leaves interpretation ambiguous, the presence of a composite decay strongly indicates an inhomogeneous system and is a useful tool for study of such systems.

Plasticized PVC provides an example of an important, frequently studied, and nominally homogeneous material that has yet to be fully characterized. X-ray, infrared and earlier NMR examinations of PVC[13-16] have been interpreted as indicating an inhomogeneous material.

Data has been discussed in terms of crystallinity or paracrystalinity, without convincing everyone that this is the correct terminology. Recent heat capacity measurements on plasticized PVC by H. Bair[17] show, among other things, two glass transitions, and raise a challenge to NMR. At 25°C the free induction signals of unplasticized PVC, PVC(0), and PVC plasticized with 17 weight percent plasticizer,[18] PVC(17), show clear differences with PVC(17) apparently exhibiting two components (figure 8). The rapidly decaying component of the PVC(17) is very nearly the same as the initial part of the PVC(0) decay. On raising the temperature to 75°C it becomes apparent from the data shown in figure 9 that PVC(5) and PVC(1) also have composite signals and appear to have three components. The longest component is expected to be plasticizer and indeed this component extrapolates to the correct relative intensity for this identification. The shortest component is little changed from sample-to-sample and seems to be

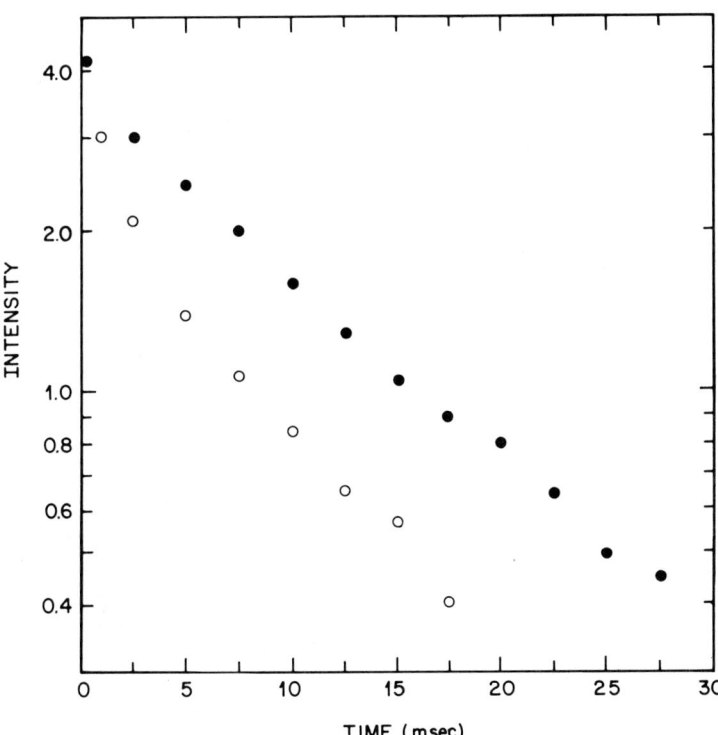

Figure 6. $T_{1\rho}$ decay of 75/25 PSAN/PMMA for: (●) $H_1 = 10G$ and (○) $H_1 = 2G$ at $-100°C$ (11)

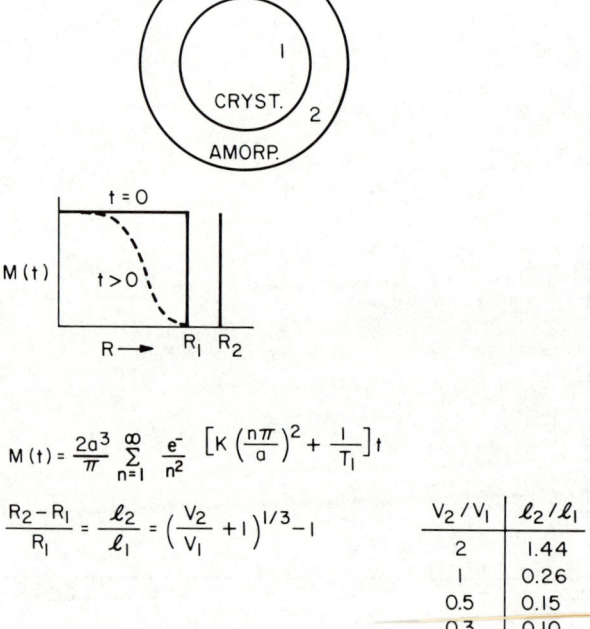

Figure 7. Model calculation for spin diffusion to a sink. If Region 2 is a sink 60% of the intensity associated with Region 1 is in the component with the largest decay time.

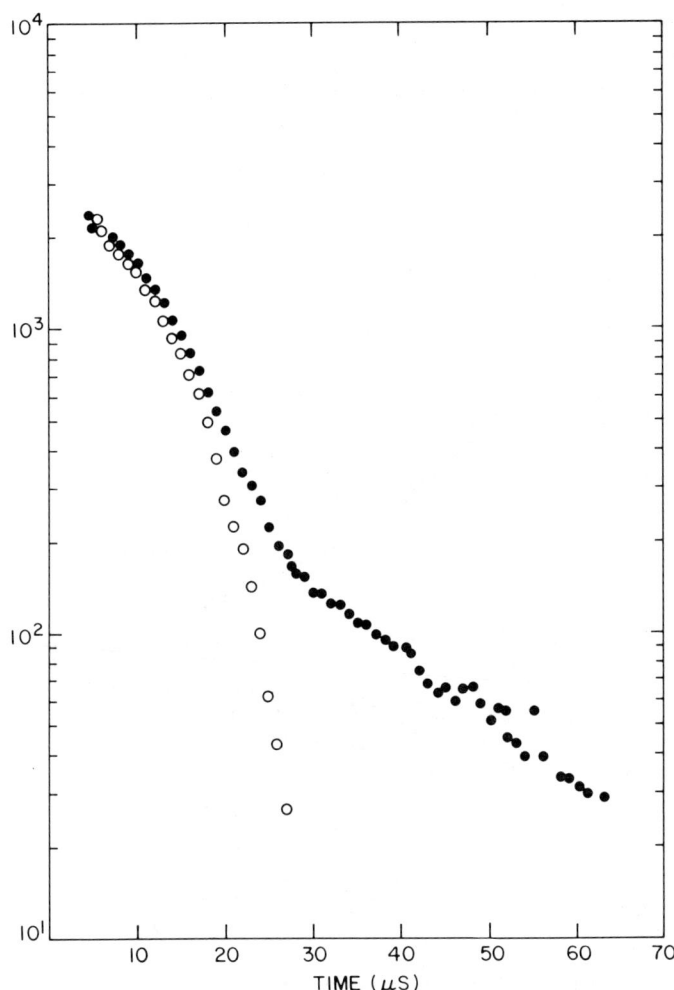

Figure 8. Free induction signals of pure and plasticized PVC at 25°C: (●) PVC (17) and (○) PVC (0)

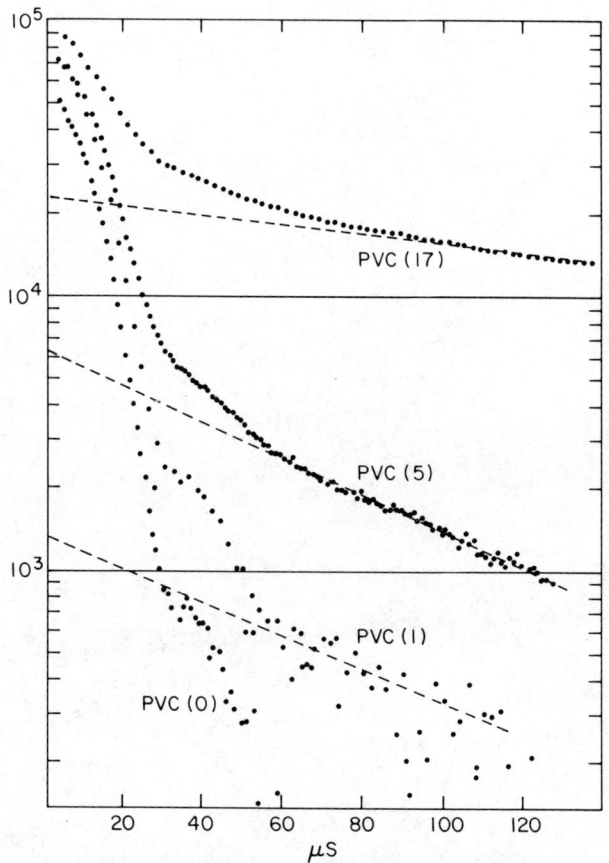

Figure 9. Free induction signals of plasticized PVC at 75°C. The numbers in parentheses are weight percent diisodecylphthalate.

unplasticized PVC. The intermediate component is then consistently interpreted as plasticized PVC. Interpretation of the free induction signal shape as resulting from a superposition of signals is supported by the observation of composite $T_{1\rho}$ decay plots, as illustrated by PVC(17) at 50°C in figure 10.

Figure 11 shows the behavior of T_2 and $T_{1\rho}$ as a function of temperature for PVC(0), (5) and (17). Resolution of the $T_{1\rho}$ curves into two component decays is arbitrary to a degree, but provides a means of characterizing the composite character that is consistent with experimental precision. Despite the scatter, largely introduced by resolution of the decays when one component has small intensity, several general conclusions can be easily be drawn from this data. First, there are two $T_{1\rho}$ minima and each $T_{1\rho}$ component shows each minimum to some degree. This again supports the contention that the plasticizer is not uniformly incorporated into the PVC and suggests that spin diffusion is coupling regions of "small" size. Second, there remains a short component in T_2, with decay time nearly the same as that of PVC(0), in PVC(17) up to temperature (120°C) where the PVC(0) T_2 increases. Third, at temperatures above 100°C the intermediate (plasticized) material has the shortest relaxation time and is acting as a sink for spins in the plasticized PVC. The data shown in figure 12 makes this last conclusion more evident and serves to illustrate a widely applicable use of NMR to identify which component, mobile or immobile, is associated with a given relaxation process. The free induction signals following the long $T_{1\rho}$ pulse are shown for three pulse durations. For the 50 μsec. pulse the FID shows the short, intermediate, and, at long times, off the plot, the long components. The FID following the 5 msec. pulse shows only the longest (plasticizer) and immobile component. The longest component is in all probability largely uncoupled from the plasticized component by virtue its large T_2.

Before leaving the subject of PVC we consider an experiment designed to indicate the size of the inhomogeneities. The Goldman-Shen[6] experiment employs a pulse sequence, shown in figure 13, that selectively heats spins associated with the short component of the FID. The spin temperature concept can be used to visualize this experiment. When the short component of the FID following the initial 90° pulse is substantially decayed, a second, -90°, pulse is applied. Magnetization associated with the longer T_2 component is returned to its initial, equilibrium configuration, along the static field and is little influenced by the two 90° pulses. The magnetization associated with the shorter T_2 component is largely destroyed and after a time long compared to T_2 it is legitimate to say that the spins in this component have been heated to a very high temperature. Subsequent internal equilibration of the spins is monitored by a third 90° pulse. The data of figure 13 is obtained by subtracting the short time monitor FID, recorded

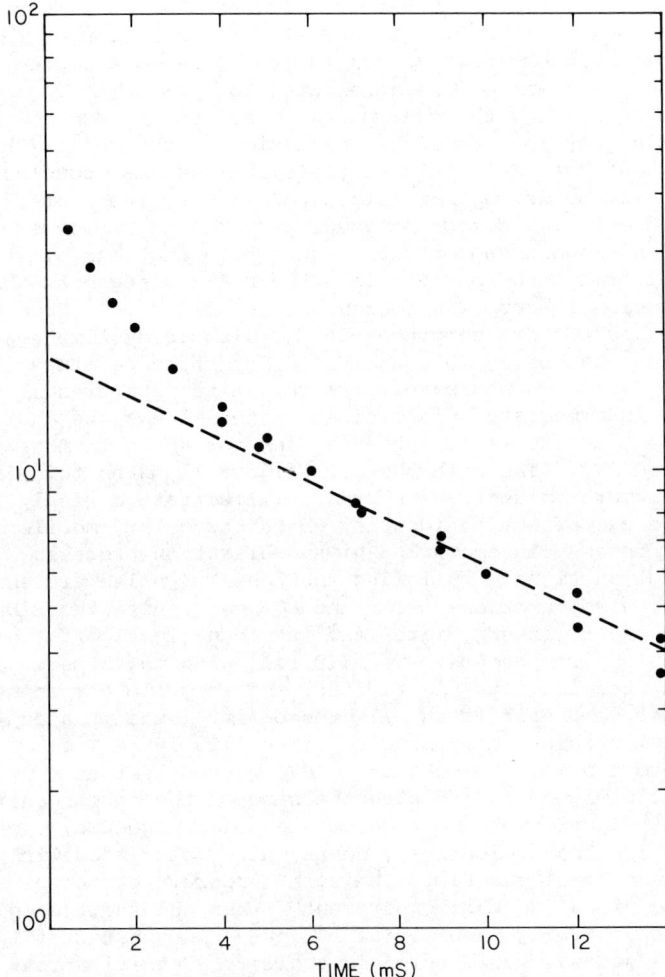

Figure 10. Free induction signal illustrating composite character: $T_{1\rho}$ *decay; PVC (17);* $H_1 = 10G$; $50°C$

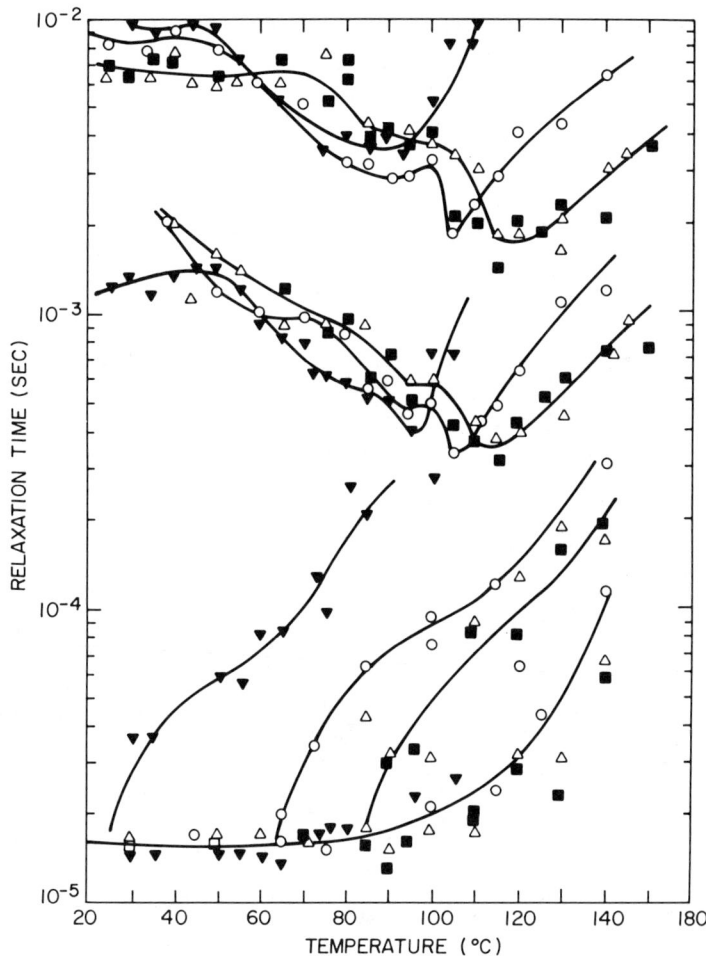

Figure 11. $T_{1\rho}$ ($H_1 = 10$ G) and T_2 data for plasticized PVC. The numbers in parentheses are weight percent diisodecylphthalate: (■), PVC (0); (△), PVC (1); (○), PVC (5); (▼), PVC (17).

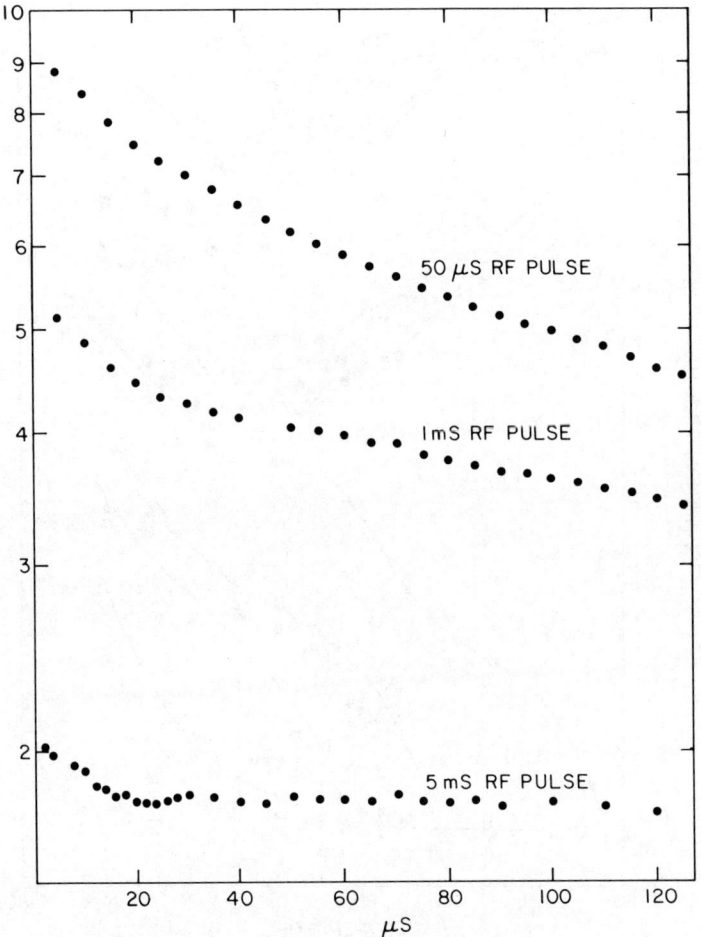

Figure 12. Free induction signals subsequent to 50 μs, 1 ms, and 5 ms $T_{1\rho}$ — RF pulses. $T_{1\rho}$ FID; PVC (17); 100°C.

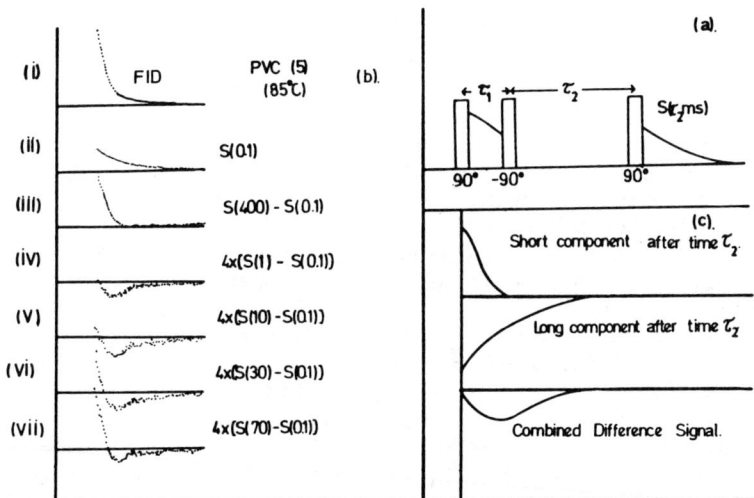

Figure 13. (a) The Goldman–Shen pulse sequence. The decay following the third 90° pulse is labeled S(τ). The spacing $\tau_1 = 30$ μs throughout. (b) Diagrams illustrating the change in shape of the decay curve in the $T_{1\rho}$ experiment on PVC (5) at $+85°C$; (iii)–(vii) show the behavior of the difference signal $S(\tau_2) - S(0.1)$. Note that (ii) and (iii) represent the mobile and rigid components, respectively. (c) Diagram presenting the difference signal in terms of the short and long component change in time τ_2.

before internal equilibration has advanced appreciably, from the monitor FID at longer times. Because equilibration takes place between the spins and lattice as well as internally, it is desirable to be able to follow each of these processes. Since the initial amplitude of the difference signal is not altered by the internal process it provides the monitor of the spin lattice relaxation; the change in shape of the difference signal shows internal relaxation; as indicated by the resolved schematic signals in figure 13. The salient result of the Goldman-Shen experiment of PVC(5) at 110°C is that internal relaxation is nearly complete in a few milliseconds, which in turn implies that immobile regions are very small, exist on a time scale longer than T_2 (20 μsec.), and probably longer than $T_{1\rho}$ (millisecs).

Figure 14 shows T_1, T_2 and $T_{1\rho}$ data[19] for PVF_2 (Kynar 821) as a function of temperature. The data are typical of a semicrystalline material and, in this case, the material shows an amorphous α transition near 0°C. Thus far we have considered only energy flow within a spin system. PVF_2 offers an opportunity to observe energy exchange between the proton and fluorine spin systems. This is interesting because of its bearing on questions regarding the homogeneity of PVF_2-PMMA blends.[20] Like spin diffusion, this exchange of energy results from dipolar interaction and hence takes place at an appreciable rate only if the two different spins are neighbors. Unlike spin-diffusion this exchange does not conserve energy and does not take place in a rigid lattice, unless the two absorption lines have overlapping tails. In the presence of molecular motion, the lattice may make up the energy unbalance and the exchange can occur. This is illustrated in figure 15 where the cross relaxation rate, σ, measured with a Solomon type pulse sequence,[21,22] is given as a function of temperature. Molecular motions associated with the α-transition have an average frequency near the nuclear difference frequency near 15°C and allow substantial cross-relaxation at this temperature. Above and below this temperature the average rate of molecular motion does not match the difference frequency and cross-relaxation is reduced. Figure 16 shows analogous data for a PVF_2-PMMA blend. In brief, the increased cross-relaxation rate near -40°C, where the average frequency for the PMMA-α-methyl group approximates the nuclear difference frequency, is taken to mean that either the fluorine nuclei have large numbers of these methyl groups as near neighbors or that the molecular motion of the PVF_2 molecules is substantially altered in the blend by the presence of PMMA molecules. In either case, the conclusion reached is that the blend is indeed homogeneous on a molecular scale in a substantial part of the amorphous regions.

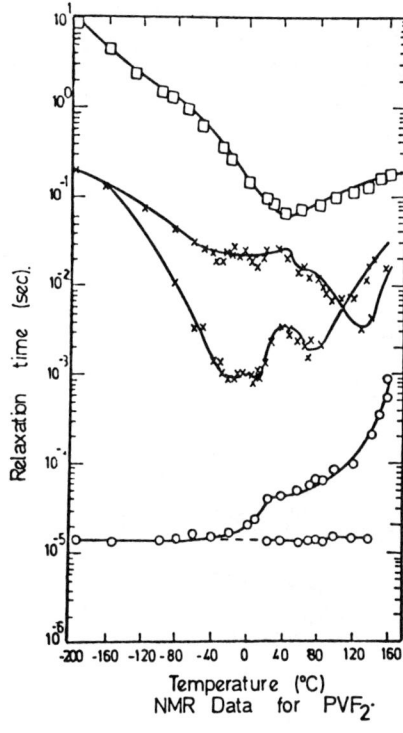

NMR Data for PVF$_2$.

Journal of Polymer Science, Polymer Physics Edition

Figure 14. T_1, T_2, and $T_{1\rho}$ data for PVF_2 as a function of temperature for protons (24): ($\square$), T_1; ($\bigcirc$), T_2; (X), $T_{1\rho}$ (H = 12.5 G)

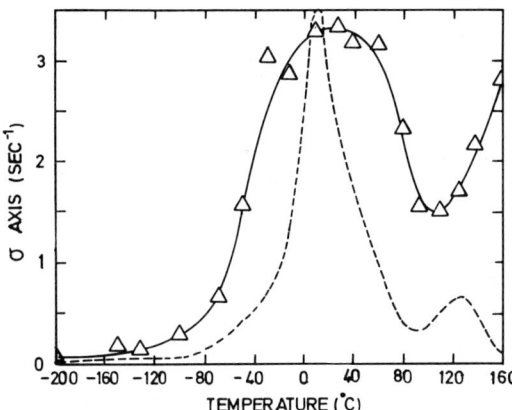

Macromolecules

Figure 15. Cross relaxation rate, σ, for PVF_2 (– – –) and a 40/60 PVF_2/PMMA blend (———) (20)

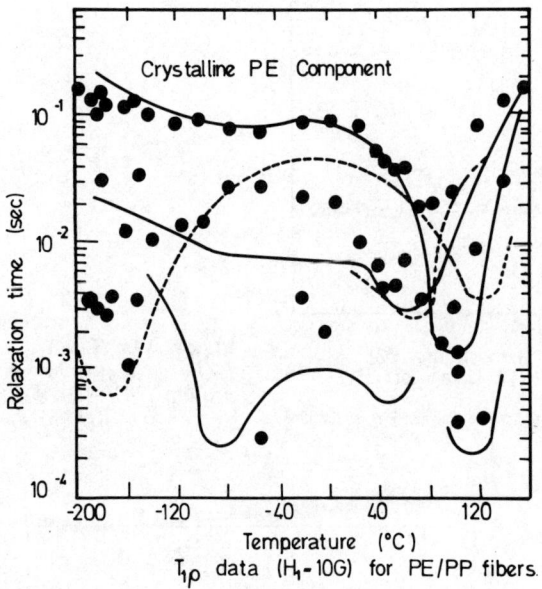

Figure 16. Comparison of T_{1p} data for blended fibers with pure PE (——) and pure PP (– – –) data (23)

A contrasting example, where NMR supplies evidence that the components are not mixed, is provided by PE, PP blended fibers prepared by a "surface growth" technique.[23] Figure 16 shows $T_{1\rho}$ data as a function of temperature for the fiber blends and the pure components. The plot appears to exhibit a large amount of scatter in the data, again resulting from resolution of $T_{1\rho}$ decay curves into more than one component. However, in the most important temperature regions, near minima, where one component dominates the relaxation behavior this resolution is less difficult and it becomes evident that the blend data closely corresponds to a superposition of the uncoupled, or slightly coupled, relaxation data for the individual components.

Literature Cited

1. Nishi, T., Kwei, T. K. and Wang, T. T., J. Appl. Phys., 1975, 46, 4157.
2. Nishi, T., Wang, T. T. and Kwei, T. K., Macromolecules, 1975, 8, 227.
3. Wang, T. T. and Sharpe, L. H., J. Adhesion, 1969, 1, 69.
4. Slichter, C. P., "Principles of Magnetic Resonance," Springer-Verlag, Berlin, 1978, p. 188.
5. Goldman, M., "Spin Temperature and Nuclear Magnetic Resonance in Solids," Clarendon Press, Oxford, 1970.
6. Goldman, M. and Shen, L., Phys. Rev., 1966, 144, 321.
7. Assink, R. A., Macromolecules, 1978, 11, 1233.
8. Robertson, J. M., "Organic Crystals and Molecules," Cornell U. Press, 1953.
9. Douglass, D. C. and Jones, G. P., J. Chem. Phys., 1966, 45, 956.
10. Naito, K., Johnson, G. E., Allara, D. L. and Kwei, T. K., Macromolecules, 1978, 11, 1260.
11. McBrierty, V. J., Douglass, D. C. and Kwei, T. K., Macromolecules, 1978, 11, 1265.
12. Garroway, A. W., comment subsequent to ACS Symposium talk.
13. Lemstra, P. J., Keller, A. and Cudby, M., J. Polym. Sci., 1978, 16, 1507.
14. McCall, D. W. and Falcone, D. R., Trans, Faraday Soc., 1970, 66, 262.
15. Davis, D. D. and Slichter, W. P., Macromolecules, 1973, 6, 728.
16. Reding, F. P., Walter, E. R., Welch, F. J., J. Polym. Sci., 1962, 56, 225.
17. Bair, H., Private communication.
18. Material used in GEON 103EP, MN = 195,000, MN = 64,000. Plasticizer is diisodecylphthalate. Samples also contain 5% lead phthalate stabilizer.
19. McBrierty, V. J. and Douglass, D. C., Macromolecules, 1977, 10, 855.

20. Douglass, D. C. and McBrierty, V. J., <u>Macromolecules</u>, 1978, 11, 766.
21. Solomon, I., <u>Phys. Rev.</u>, 1955, 99, 559.
22. Solomon, I and Bloembergen, N., <u>J. Chem. Phys.</u>, 1955, 25, 261.
23. McBrierty, V. J., Douglass, D. C. and Barham, P. J., submitted to J. Polym. Sci.
24. McBrierty, V. J., Douglass, D. C., Webber, T. A., Journal of Polymer Science, Polymer Physics Edition, 1976, 14, 1271.

RECEIVED June 18, 1980.

Multiple-Pulse NMR of Solid Polymers: Dynamics of Polytetrafluoroethylene

A. D. ENGLISH and A. J. VEGA

E. I. Du Pont de Nemours & Co., Inc., Central Research and Development Department, Experimental Station, Wilmington, DE 19898

Previous studies of polytetrafluoroethylene (PTFE) by conventional[1-5] and multiple pulse NMR[6-8] have given some insight into the different types of polymer motion that are present in both crystalline and amorphous regions. Multiple pulse NMR studies have determined that ^{19}F chemical shift spectra of fluoropolymers can be obtained[6], that the principal values of the chemical shift tensor can be determined in a randomly oriented sample[8], and additionally that the orientation of the chemical shift tensor with respect to a molecular axis system can be determined for an oriented sample.[7] However, these previous studies did not address themselves to the use of multiple pulse NMR to determine crystallinity or to observe various motions and/or relaxations. We have recently shown[9] that ^{19}F chemical shift spectra obtained using multiple pulse techniques can be used to determine crystallinity and to directly observe general types of macromolecular motions in polycrystalline PTFE. Our present results demonstrate that chemical shift spectra obtained using multiple pulse techniques can be used to obtain well defined values of polymer crystallinity that correlate well with other measurements and when combined with relaxation measurements can be used to <u>directly detect macromolecular motions in both crystalline and amorphous regions in randomly oriented samples.</u>

Results

Variable temperature (-150° to +350°) chemical shift spectra have been obtained for melt-recrystallized samples of PTFE. The NMR spectra were acquired using the REV-8 sequence[15] which consists of a long train of closely spaced, high power rf pulses of fixed width, and varying in phase in a cyclical fashion. The cycle consists of eight π/2 pulses. The total cycle time used was 43.2 μsec with a shortest spacing between pulses of 3.6 μsec. The nominal resonance frequency was 84.6

0-8412-0594-9/80/47-142-169$05.75/0
© 1980 American Chemical Society

MHz, a π/2 pulse was 2.2 μsec and the data were acquired with the carrier frequency below resonance (the rhs of the spectrum in each figure). These spectra may be used to determine polymer crystallinity and to determine various kinds of macromolecular motion. Also spin-lattice relaxation times in the rf interaction frame (T_{1xz}^{REV}) in addition to conventional T_1 and $T_{1\rho}$ relaxation times have been measured to help elucidate the various mechanisms responsible for the observed chemical shift line shapes.

Chemical shift spectra of PTFE obtained at 259° are shown in Figure 1. These lineshapes, for three different samples of varying crystallinity, may be seen to be a linear combination of two lineshapes: one is characteristic of an axially symmetric powder pattern and the other of an isotropic chemical shift tensor. At this temperature these two lineshapes differ greatly and may be numerically decomposed.

In Figure 2, these lineshapes are decomposed into the isotropic lineshape and the axially symmetric powder pattern lineshape. This same procedure was used for eight samples. The narrow component corresponds to the amorphous fraction of the polymer and the axially symmetric powder pattern to the crystalline fraction of the polymer. Moreover, the amorphous as well as the crystalline line shapes extracted from samples of different crystallinity are essentially identical, which indicates that the structure and dynamics in the crystalline and amorphous fractions are independent of the degree of crystallinity.

The relative contributions to the intensity of a composite line shape are thus a measure of the degree of crystallinity of the PTFE sample. The crystallinities thus determined are compared to values determined by density[10,11] measurements in Figure 3. These results indicate an excellent correlation between crystallinities measured with the two techniques; however, the agreement is not absolute, which is not unusual when comparing values of crystallinity measured by different techniques which use different criteria for crystallinity[12].

Decomposed chemical shift spectra 29 of melt-recrystallized PTFE at three temperatures (Figure 4) illustrate that both the crystalline and amorphous regions of the polymer undergo dramatic lineshape changes over this temperature range. The amorphous lineshape (left side of Figure 4) is completely asymmetric at the lowest temperature and at sufficiently high temperatures becomes symmetric; the crystalline lineshape is also asymmetric at the lowest temperature but retains axial symmetry to at least 259. In fact, the crystalline lineshape remains axially symmetric to very near the melting point (327°). Additionally, the crystalline and amorphous lineshapes have very nearly the same lineshape at the lowest temperature (the lineshapes are invariant to temperature below -128) indicating

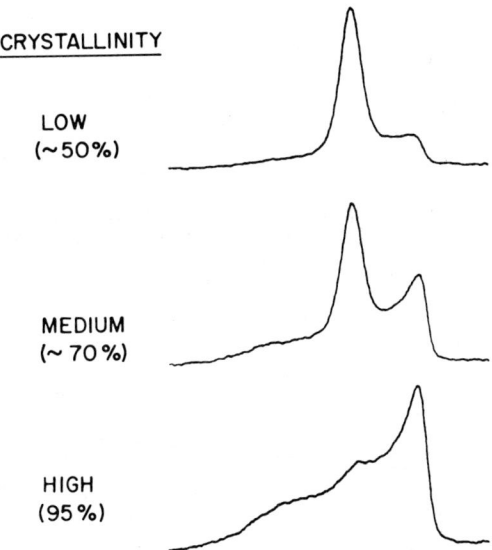

Figure 1. F-19 REV-8 chemical shift spectra of PTFE samples of varying crystallinity obtained at 259°C. The total width of the spectra is about 17 kHz.

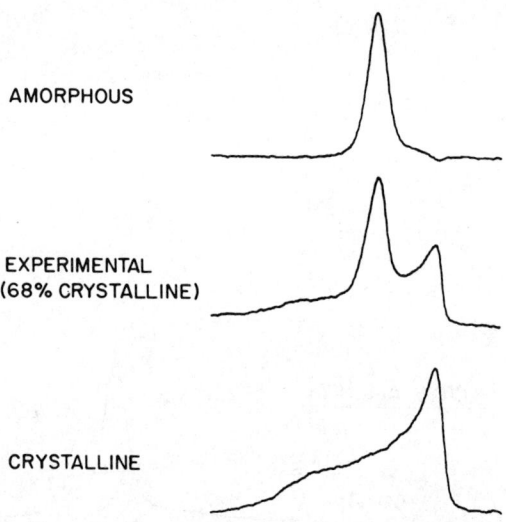

Figure 2. F-19 REV-8 chemical shift lineshapes of a 68% crystalline PTFE sample at 259°C and decomposed lineshape of the amorphous and crystalline fractions

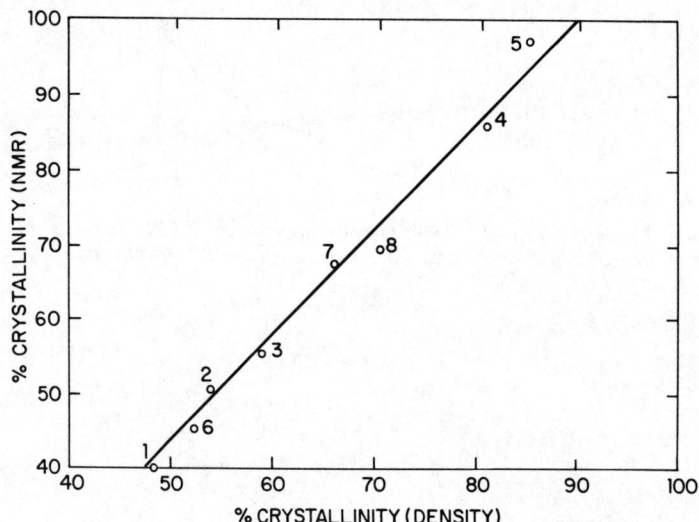

Figure 3. Crystallinities of several melt-processed and annealed-virgin PTFE samples as determined from decomposition of chemical shift lineshapes vs. crystallinities determined from density measurements

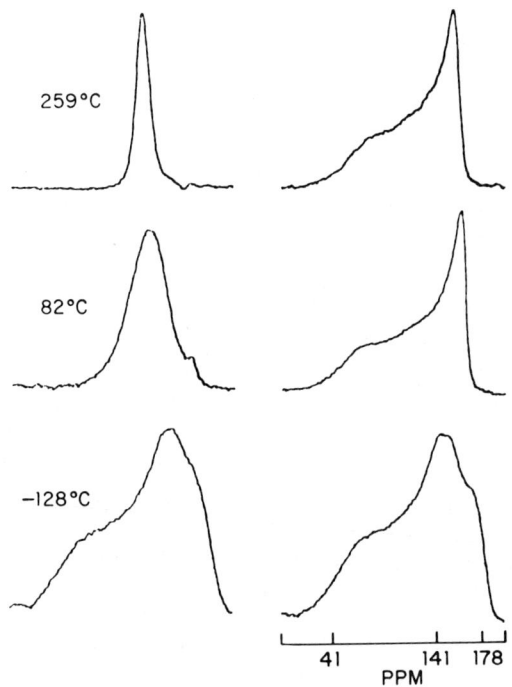

Figure 4. Decomposed F-19 REV-8 chemical shift spectra of the amorphous (left) and crystalline (right) fractions of PTFE at various temperatures

that in the slow motion limit the environments of the amorphous and crystalline regions are essentially indistinguishable. The figure also gives the 3 principal components of the chemical shift tensor as σ_{xx} = 178 ppm, σ_{yy} = 141 ppm, σ_{zz} = 41 ppm with respect to $CFCl_3$.

A more complete set of decomposed lineshapes (Figure 5) allows us to make more detailed conclusions about the macromolecular motions in the crystalline and amorphous regions. The crystalline lineshape is seen to retain the asymmetric lineshape until ∿10° and to reach a completely axially symmetric lineshape by ∿30°. The crystalline lineshapes are amenable to a detailed study of the rate at which the lineshape changes from asymmetric to axially symmetric. Figure 5 shows small temperature increments during which the crystalline lineshape undergoes this transformation. The temperature dependence of these lineshapes can be simulated by a model of rotational diffusion about the chain axis and these calculated spectra are also shown in Figure 6. The temperature dependence of the rate of this process is shown as an Arrhenius plot in Figure 7 and the calculated least squares fitted activation energy is 48±11 kcal/mole (95% confidence limits).

The amorphous lineshape becomes axially symmetric by -60° and then begins to collapse into a symmetric lineshape. The transformation of the amorphous lineshape from asymmetric to axially symmetric is imprecisely defined because the process which transforms it into a symmetric lineshape becomes dominant before the lineshape becomes completely axially symmetric. Figure 8 shows in detail representative amorphous chemical shift lineshapes as a function of temperature.

Figure 9 gives values of $T_{1\rho}$ and T_{1xz}^{REV} for one of the samples as a function of temperature. Both the $T_{1\rho}$ and T_{1xz}^{REV} data show two components in the relaxation data below 30°C. Above 30°C the relaxation decays are singly exponential. The fractions with the long and short relaxation times represent the crystalline and amorphous phases, respectively.

The amorphous T_{1xz}^{REV} and $T_{1\rho}$ minima occur at approximately the same temperature where the amorphous lineshape narrows into an axially symmetric pattern, indicating that these phenomena are related to the same macromolecular process. The crystalline T_{1xz}^{REV} and $T_{1\rho}$ relaxation times approach a minimum at 30°C, which is clearly related to the lineshape changes shown in Figure 6.

The single exponential T_{1xz}^{REV} and $T_{1\rho}$ decays above the first order phase transition at 30°C are not accompanied by related chemical-shift lineshape changes: the crystalline line shape does not change and, as we will explain shortly, the narrowing of the amorphous lineshape at high temperature cannot be directly related to the observed relaxation times. Most peculiar is the fact that, although the crystalline and amorphous line shapes are significantly different at higher temperatures, the relaxation times in the rf interaction frame are identical in

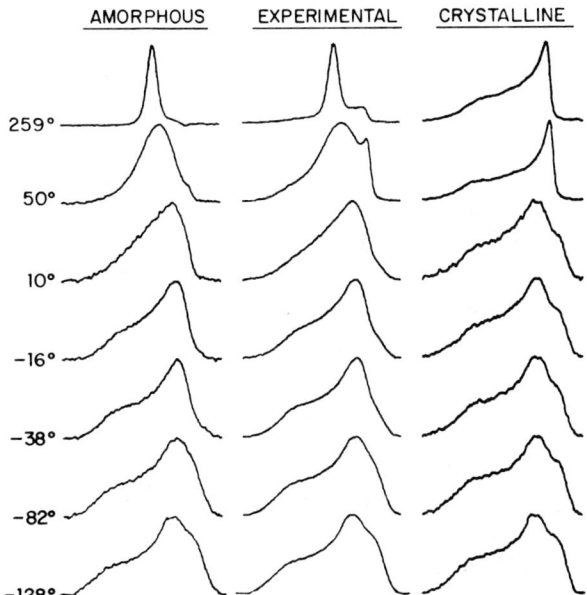

Figure 5. F-19 REV-8 chemical shift lineshapes of melt recrystallized PTFE at various temperatures. The experimental spectra (center) are of a 68% crystalline sample. The amorphous (left) and crystalline (right) lineshapes were obtained by decomposition of experimental spectra.

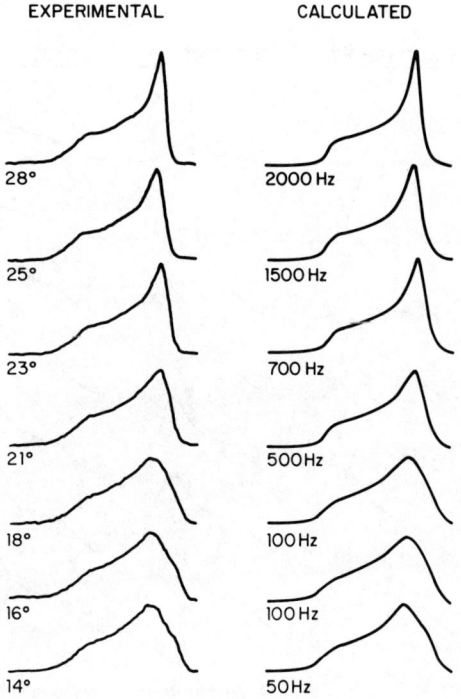

Figure 6. *Experimental F-19 REV-8 chemical shift lineshapes of the crystalline fraction of melt-crystallized PTFE between 14° and 28°C (left) and calculated lineshapes for rotational diffusion about the chain axis with the diffusion coefficient as indicated (right)*

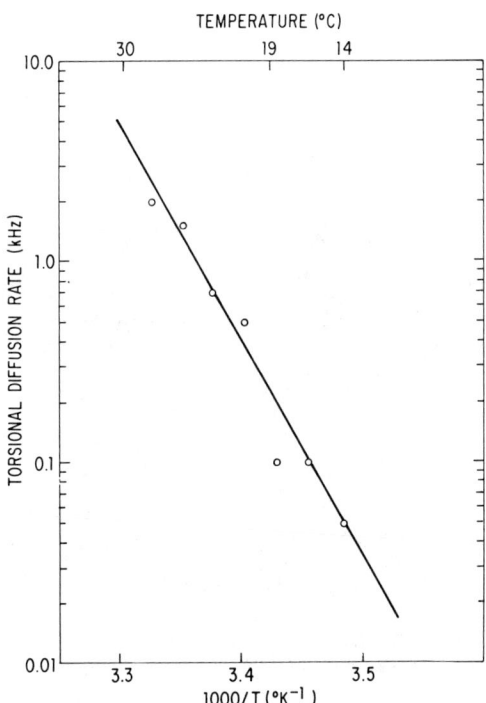

Figure 7. Temperature dependence of the rate of rotational diffusion about the chain axis of PTFE molecules in the crystalline regions of melt-recrystallized samples. $E_a = 48 \pm 11$ *kcal/mol.*

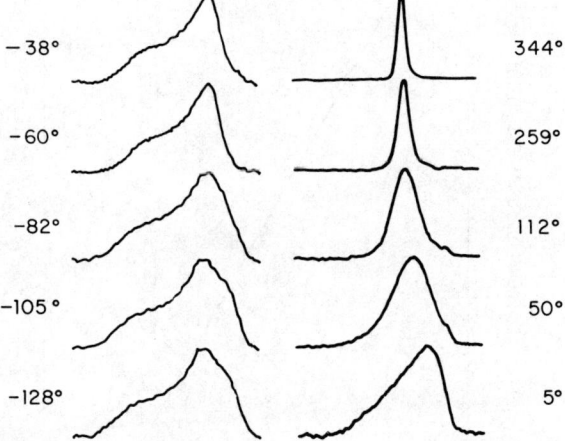

Figure 8. F-19 REV-8 chemical shift lineshapes of the amorphous fraction of melt-recrystallized PTFE at various temperatures

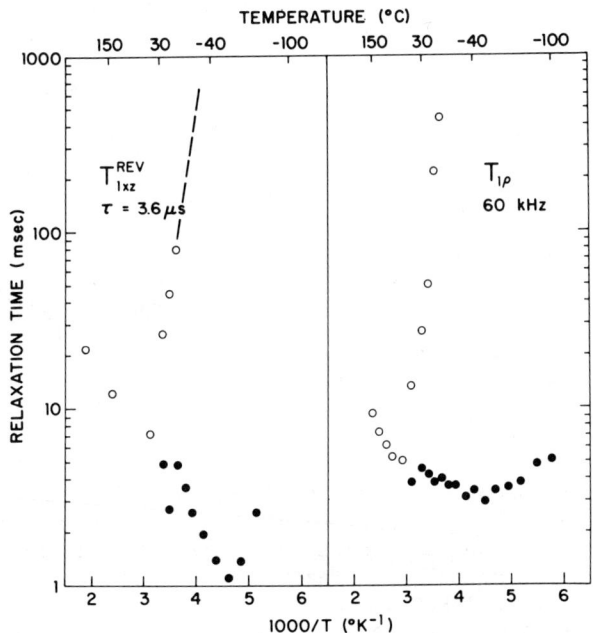

Figure 9. Temperature dependence of the $T_{1\rho}$ and T_{1xz}^{REV} relaxation times of a 68% crystalline melt-recrystallized PTFE sample: (●) denote the short-relaxation-time component in a double exponential decay. T_{1xz}^{REV} times longer than 100 ms and $T_{1\rho}$ times longer than 1 s are not shown since their experimental values could not be determined accurately. The expected T_{1xz}^{REV} temperature dependence below 15°C is indicated by (– – –).

the two phases. This probably indicates that the same dynamical process dominates the relaxation rate in both phases.

The lineshape of the amorphous fraction (shown in Figure 8) is seen to be changing from that of a chemical shift tensor of axial symmetry at -38°C to one with spherical symmetry near the melting point. Hence, the anisotropic reorientation at low temperatures gradually changes into an essentially isotropic reorientation at high temperatures. As is often the case with magnetic resonance lineshape phenomena of this kind, one might suppose that this isotropic motion already exists at a slow rate at the lowest temperatures and that the increase of the rate with increasing temperature is reflected by the variations in the lineshape. In the case of the amorphous lineshapes, this would imply that above 50°C the lineshapes were Lorentzian and <u>homogeneously</u> broadened (fast motion limit). However, this would also imply that the relaxation time T_{1xz}^{REV} be equal to the inverse half width of the Lorentzian line. Since in our case T_{1xz}^{REV} is two orders of magnitude longer than the inverse line width, we must conclude that the lineshapes are <u>inhomogeneously</u> broadened. The correct description of the amorphous lineshapes is that reorientational motion takes place at a rate fast compared to the chemical shift range in question (10 kHz) and that the angular <u>amplitude</u> rather than the <u>rate</u> of this motion grows with increasing temperature until essentially isotropic motion is achieved at the melting point.

Discussion

PTFE is known to undergo two crystalline first-order transitions at 19° and 30° and three viscoelastic relaxations have been observed. The temperature dependence of the relaxations (α, β, γ) as determined by dielectric and anelastic measurements is shown in the relaxation map of Figure 10 along with the temperatures of the crystalline phase transitions (vertical dotted lines).[13,14,15] Also indicated in Figure 10 are the results of rotational diffusion rates obtained in this study. The γ relaxation is thought to occur in the amorphous regions and has an activation energy E_a = 18 kcal/mole.[10] The β relaxation occurs in the crystalline regions and is related to reorientation about the chain axis and has E_a = 34 kcal/mole.[10] The α relaxation has an E_a = 88 kcal/mole[10] and may involve the crystalline and/or amorphous regions.

The chemical shift lineshape changes observed for the crystalline fraction of melt-recrystallized PTFE from $\sim$14°C to $\sim$28°C (Figure 6) are due to reorientation about an axis essentially parallel to σ_{zz} (which is tilted $\sim$12° from the molecular chain axis). The temperature dependence of the lineshape changes can be simulated as an increase in the rate of rotational diffusion about this axis (Figure 6) and this process has an activation energy of 48±11 kcal/mole. In addition, the

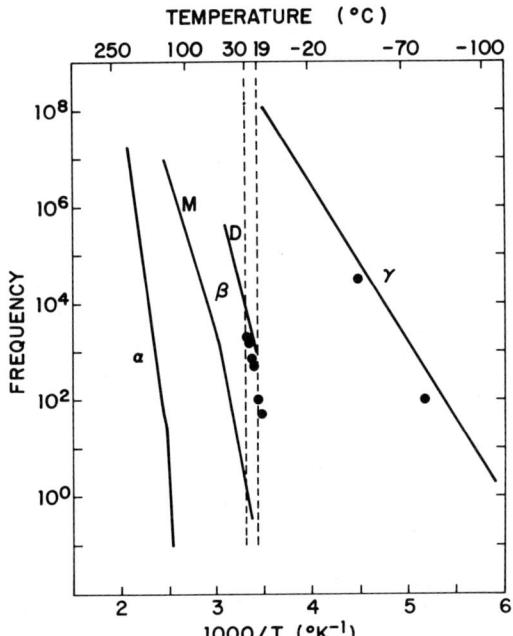

Figure 10. Relaxation map of PTFE.

The full lines indicate the approximate temperature dependence of the α, β, and γ relaxation frequencies determined by dynamic mechanical and dielectric measurements (4, 21) (———). Two branches have been observed in the β relaxations: M(dynamic mechanical and dielectric data) and D(dielectric data only). Crystalline phase transitions are represented by (– – –); the results of this work (●). The points near 19°C represent the rotational diffusion rate of the molecules in the crystalline regions (see Figure 9). The points at −50° and −80°C are based on the T_{1xz}^{REV} and $T_{1\rho}$ minima and the amorphous chemical shift lineshapes, respectively (see text).

$T_{1\rho}$ and T_{1xz}^{REV} relaxation data show minima near 30°C where the rotational diffusion rate is approximately equal to the effective spin locking field. These three facts indicate that the process responsible for the lineshape changes in crystalline regions is a relaxation process that can be described as a reorientation about the chain axis that increases in rate with increasing temperature. This process is clearly the β relaxation. The relaxation rates determined by NMR are in good agreement with some of the previous dielectric results (Figure 10). The activation energy found from these NMR results is also in good agreement with some of the dielectric results but disagrees with the often quoted value of 34 kcal/mole.

The lineshape changes observed for the amorphous fraction of the polymer between -128° and -60° (Figure 8) indicate some type of relaxation whose origin is similar to that just described. It appears that this process involves reorientation about the local chain axis and the rate of this process is $\sim$ 0.1 kHz at $\sim$ -80°C. The relaxation data of Figure 9 are consistent with this explanation and indicate that the rate of this motion is $\sim$ 30 kHz at -50°C. However, this process is coupled with a higher temperature process (see below) and does not give definitive lineshapes which are amenable to lineshape analysis. We can estimate that this relaxation has an activation energy of 16±5 kcal/mole. Therefore, we conclude that the process responsible for these results is the γ relaxation.

The relaxation times shown in Figure 9 indicate that from -128° to 300° only two minima are observed. These two minima are associated with the γ and β relaxations previously observed in anelastic and dielectric measurements. We see no evidence of any other relaxations on the time scale of $T_{1\rho}$ and T_{1xz}^{REV} (10-100 kHz). Therefore, these NMR epxeriments as well as conventional NMR experiments[4] do not detect the α relaxation. Furthermore, the reorientational motion of the amorphous chain direction (see below) cannot be identified with the α relaxation since this motion involves rates faster than 100 MHz below room temperature while the α relaxation is slower than 1 Hz below 100°C. The absence of any evidence for the α relaxation in these NMR experiments indicates that the α relaxation does not correspond to the motions associated with the β or γ relaxations or chain axis reorientation in either the crystalline regions or the amorphous regions (see below), but may be related to some type of chain translation in either the amorphous or crystalline regions.

It is clear from the chemical shift spectra that the only process observable in the crystalline regions from -128° to +320° (near the melting point) is reorientation about the chain axis. Translational displacements along the chain axis in well-ordered crystalline regions (which have been previously observed[13-15]) would not give rise to any observable change in the chemical shift spectra. However, the chemical shift spectra of the amorphous regions indicate that considerable motion occurs by $\sim$ -40° that is not seen by the relaxation measurements ($T_{1\rho}$, T_{1xz}^{REV} and T_1) and is not identifiable with the α, β, or γ relaxations. The molecular motion responsible for the observed lineshape changes in the amorphous regions above -40° cannot be attributed to a relaxation process in which the increase of the rate of the molecular motion with temperature is responsible for the lineshape effects.

The rate of the molecular motion responsible for the observed amorphous lineshape changes above -40° must be much less than or much greater than 1-10 kHz (T_{1xz}^{REV}) and much less or much greater than 100 MHz (T_1) from -40°C to 260°C[16] since no minimum was observed in the relaxation data. In addition, the rate must be $\gtrsim$1-10 kHz to cause the observed lineshape changes. Therefore, the rate must be greater than 100 MHz. (At these temperatures, the rate of the α relaxation is <<1 Hz). These considerations lead us to conclude that the process responsible for the observed lineshape changes may be idealized as one in which the rate of the motion is fast ($\gtrsim$100 MHz) at all temperatures (-40°C to 260°C) and the amplitude of the motion grows as a function of temperature. Since the chemical shift lineshape is nearly axially symmetric at $\sim$ -68°C (due to rapid reorientation about the local chain axis), we can describe this motion whose amplitude grows with temperature as reorientation of the local chain axis.

The reorientation of the local chain axis gives rise to the partial narrowing of the amorphous REV-8 spectra and is of a random nature. In a short period of time, a particular segment of the macromolecular chain assumes a distribution of directions which deviate from its orientation at rest. In analogy to the description of molecular ordering in liquid crystals[17], we use the concept of a local order parameter for the quantitative characterization of the extent of these fluctuations.

In liquid crystals, the order parameter, S_β, is a well-defined parameter for the characterization of NMR spectra where very fast reorientational motion of the molecules gives rise to a uniform scaling (with a factor S_β) of spectral quantities such as nuclear dipolar splittings.[17-19] This uniform scaling is obtained in liquid crystals because the following three conditions are satisfied: (i) The term in the nuclear spin Hamiltonian being averaged by the fast motion is proportional to $(3\cos^2\theta-1)$, where θ is the angle between the magnetic field $\vec{H}_o$ and an instantaneous molecular vector $\vec{r}$ (See Figure 11). (ii) The distribution of the orientations sampled by the vector $\vec{r}$ during the motion is axially symmetric with respect to a local director $\vec{d}$. (iii) All the molecules in the sample reorient identically, i.e. the distribution $P(\beta)$ of the angles β between the instantaneous molecular vectors $\vec{r}$ and their local average direction $\vec{d}$ is the same for all molecules. Since the motion is very fast, an NMR spectrum is observed which is proportional to the average value $<3\cos^2\theta-1>$, and if the above conditions are fulfilled we have the relationship

$$<3\cos^2\theta-1> = S_\beta(\cos^2\theta'-1) \tag{1}$$

where $S_\beta = 1/2 <3\cos^2\beta-1>$

$$= \frac{\int 1/2(3\cos^2\beta-1)P(\beta)\sin\beta d\beta}{\int P(\beta)\sin\beta d\beta} \tag{2}$$

θ' being the angle between the director $\vec{d}$ and the magnetic field H.

In the amorphous regions of PTFE we identify $\vec{d}$ with the time average of the local chain direction and $\vec{r}$ with its instantaneous direction. Since the torsional motion about the chain axis above -68°C is such that we retain effective axial symmetry of the chemical shift with respect to the molecular chain axis, we do have the angular dependence required by condition (i):

$$\sigma = \sigma_{iso} + 1/3\Delta\sigma(3\cos^2\theta-1) \tag{3}$$

where $\sigma_{iso} = 1/3(\sigma_{||} + 2\sigma_\perp) \tag{4}$

$$\Delta\sigma = \sigma_{||} - \sigma_\perp \tag{5}$$

If the conditions (ii) and (iii) were also satisfied, we would have observed a line shape corresponding to an axially symmetric chemical shift tensor to near the melting point and we would have calculated an unambiguous value for the order parameter from

$$S_\beta(T) = \Delta\sigma(T) / \Delta\sigma(T_o), \tag{6}$$

where $\Delta\sigma(T)$ is the line width at temperature T, and T_o is the temperature at which the amplitude of the motion is effectively zero. However, as can be seen in Figure 8, the lineshape is essentially symmetric above 50°C. This is most likely due to the facts that the chain bending motion is not really symmetrically distributed about the director, and that molecules in different sites of the polymer have different amplitudes of motion, giving rise to a distribution of order parameters. Nevertheless, we feel that the observed width of the narrowed line is very closely related to an order parameter which corresponds to some kind of average value for $S_\beta = 1/2<3\cos^2\beta-1>$. Experimental values for $\Delta\sigma$ were obtained by least-squares curve fitting of the amorphous lineshapes at temperatures above -40°, to an axially symmetric chemical shift tensor. From these values S_β was calculated with Equation (6), where $\Delta\sigma(T_o)$ was taken as the $\Delta\sigma$ of the crystalline spectra above room temperature. Figure 12 gives the temperature dependence of the local

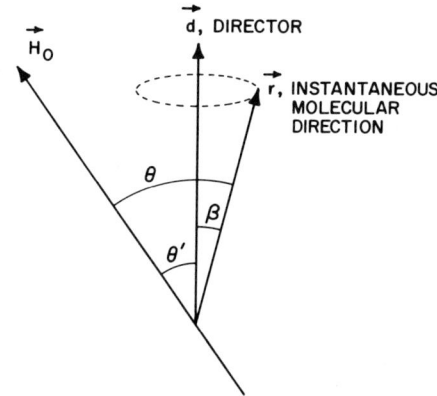

Figure 11. Definition of angles describing the relative orientations of the magnetic field ($\vec{H}$), the instantaneous molecular direction ($\vec{r}$), and the average molecular direction ($\vec{d}$) in a dynamically disoriented system. In amorphous PTFE $\vec{r}$ represents the instantaneous chain direction and $\vec{d}$ the local director.

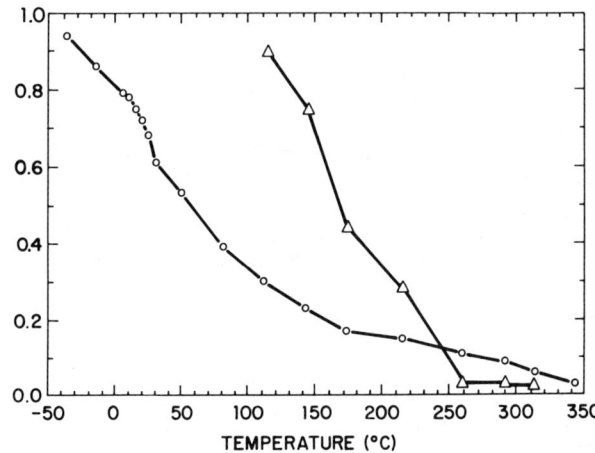

Figure 12. Temperature dependence of the local-order parameter, S_β, of the fluctuating chain directions in the amorphous regions of PTFE (○) and a TFE/HFP (△) copolymer

order parameter obtained in this fashion for the amorphous regions in PTFE and in a TFE/hexafluoropropylene copolymer.

We can now speculate as to the molecular nature of this reorientational motion of the PTFE backbone in the amorphous state. We assume that our experimentally determined order parameters closely represent the average value of $1/2<3\cos^2\beta-1>$ of all the molecular chains in the amorphous regions, i.e., we ignore a distribution of order parameters and the effects of nonaxially symmetric deviations from the local director.

Two models for the dynamic distribution of angular displacements β are considered. The first is a Gaussian distribution, $P(\beta) \propto \exp(-\beta^2/2\beta^2_0)$, which is characterized by an rms angle or standard deviation β_0. Substitution of this expression for $P(\beta)$ in Equation (2) provides the dependence of the temperature dependence of β_0.[18] These results are shown in Figure 13. It is seen that the magnitude of the angular disorder in the amorphous regions at the crystalline melting point is of the order of 60°.

The temperature dependence of the standard deviation of the fluctuations can be related to the potential energy required to introduce disorder into the polymer chain. Figure 13 shows a linear dependence of log β_0 on 1/T over a fairly large temperature range. This suggests that higher energy chain conformations are populated according to a Boltzmann distribution. From the slope of the line in Figure 13 we deduce a potential energy difference $\Delta E = 1.2$ kcal/mole for the creation of local disorder in the backbone of amorphous PTFE. A comparison of this ΔE to values predicted by the four state rotational model of Bates and Stockmayer[20] (where $t^{\pm}t^{\pm}$ is the ground state) shows reasonable agreement with their values for $t^{\pm}g^{\pm}$ defects ($\Delta E = 1.4 \pm 0.4$ kcal/mole) or $t^{\pm}t^{\mp}$ defects ($\Delta E = 1.1 \pm 0.7$ kcal/mole). In addition, this value for the potential energy difference is essentially identical to that found from infrared studies of PTFE, which postulated the existence of helix reversals ($t^{\pm}t^{\mp}$).[21-23] However, since helix reversals are unlikely to be responsible for large reorientations of the chain axis, the motion that we observe is more likely due to $t^{\pm}g^{\pm}$ defects. On the other hand, our model of a continuous Gaussian distribution is obviously not compatible with discrete conformational changes in the backbone. Instead, it may possibly be related to large amplitude anharmonic vibrations. Indeed, we have established that the macromolecular motion is of a high frequency ($>\sim 100$ MHz). This vibration could be a perturbed longitudinal acoustic mode and may be detectable with Raman spectroscopy at low temperatures where the concentration of defects may be small enough so as not to cause the Raman band to be so broad as to be unobservable.[24,25]

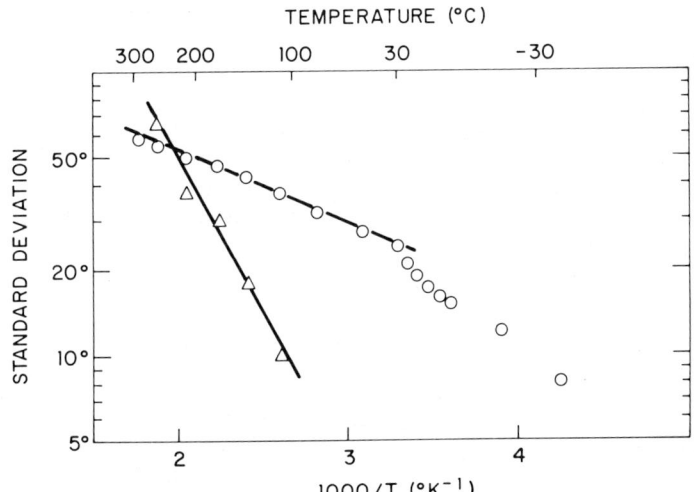

Figure 13. Temperature dependence of the rms value of a Gaussian distribution of the amplitude of chain axis fluctuations in the amorphous regions of PTFE (1.2 kcal/mol) and a TFE/HFP (5.0 kcal/mol) copolymer

Another model to explain the growing dynamic disorder with increasing temperature is directly related to the creation of gauche conformations in the predominantly trans-configuration PTFE chain. In this model, the angle β can have essentially one of two distinct values: $\beta_o = 0$ in the case of an undeformed chain, and $\beta = \beta_k$ in parts of the chain that are bent away from their original direction due to the criteria of a kink. An analysis of the PTFE molecular geometry shows that $\beta_k = 56°\pm4°$ for a trans-gauche conformational change.[7, 15] If the number of CF_2 groups in kinks is p times the number of CF_2 groups in undeformed parts of the chain, then

$$S_\beta = [1/2(3\cos^2\beta_o - 1) + p1/2(3\cos^2\beta_k - 1)]/(1+p)$$
$$\simeq 1/(1+p).$$

We again stress that this relationship has rigorous physical meaning only if the distribution of local kink orientations has cylindrical symmetry about the director; nevertheless, it may provide a semi-quantitative relationship between the REV-8 lineshape and the concentration of chain deformations if the process is at least of a highly random nature. If the ratio p changes with temperature according to a Boltzmann factor, we expect the quantity

$$(1-S_\beta)/S_\beta \simeq p \sim \exp(-\Delta E/kT)$$

to reflect this temperature dependence. Figure 14 shows a surprisingly good agreement with this model. The corresponding potential energy for the creation of chain kinks is $\Delta E = 3.3$ kcal/mole. This is more than twice the value of 1.4 kcal/mole predicted by Bates and Stockmayer[20] for the introduction of a gauche conformation and may be due to the fact that a kink requires at least two gauche conformations in order to maintain the overall chain position in the polymer matrix.

Additionally, a Reneker[26] type defect is suggested by the potential energy difference found for this second model since the calculated energy required for the formation of this type of defect in amorphous polyethylene is $\simeq 4$ kcal/mole[26] (polytetrafluoroethylene should be similar). This defect model is also attractive in that it provides a mechanism for establishing a correspondence between the chain axis translation observed in crystalline PTFE by x-ray[27] (which is consistent with our results) and a translation (longitudinal motion) of the local chain axis in the amorphous regions ("reptation").[28] Since the chain axis is "straight" in the crystalline regions but not in the amorphous regions, segmental reorientation occurs only in the latter and is reflected in our results. We find this an attractive possibility, because it explains why we observe the same $T_{1\rho}$ and T_{1xz}^{REV} relaxation times for the amorphous and the crystalline signals above 30°C.

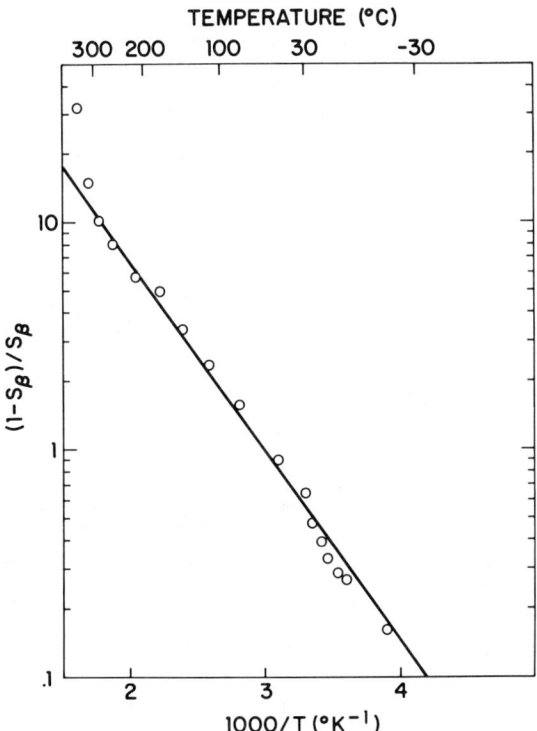

Figure 14. Temperature dependence of $p = (1 - S_\beta)/S_\beta$, S_β *being the local-order parameter in the amorphous regions of PTFE. This quantity is related to the population of kinks (*see *text).* $\triangle E = 3.3$ *kcal/mol.*

Acknowledgements

We are much indebted to the late Professor R. W. Vaughan for valuable advice and his constant enthusiasm. We gratefully acknowledge discussions with Dr. C. A. Sperati, Dr. H. W. Starkweather, and the skilled technical assistance of R. O. Balback.

References

1. C. W. Wilson and G. E. Pake, J. Chem. Phys. 27, 115 (1957).
2. W. P. Slichter, J. Polym. Sci. 24, 173 (1957).
3. D. Hyndman and G. F. Origlio, J. Appl. Phys. 31, 1849 (1960).
4. D. W. McCall, D. C. Douglass, and D. R. Falcone, J. Phys. Chem. 71, 998 (1967).
5. V. J. McBrierty, D. W. McCall, D. C. Douglass, and D. R. Falcone, Macromolecules 4, 584 (1971).
6. D. Ellett, U. Haeberlen, and J. S. Waugh, J. Polym. Sci., Polym. Lett. Ed., 7, 71 (1969).
7. A. N. Garroway, D. C. Stalker, and P. Mansfield, Polymer 16, 171 (1975).
8. M. Mehring, R. G. Griffin, and J. S. Waugh, J. Chem. Phys. 55, 746 (1971).
   a) J. Schaefer, E. O. Stejskal, and R. Buchdahl, Macromolecules 8, 291 (1975).
   b) J. Schaefer, E. O. Stejskal, and R. Buchdahl, Macromolecules 10, 384 (1977).
9. A. D. English and A. J. Vega, Macromolecules 12, 353 (1979).
10. R. K. Eby and K. M. Sinnott, J. Appl. Phys. 32, 1765 (1961).
11. C. A. Sperati and H. W. Starkweather, Fortschr. Hochpolym. Forsch. 2, 465 (1961).
12. A. L. Ryland, J. Chem. Educ. 35, 80 (1958).
13. N. G. McCrum, B. E. Read, and G. Williams, Anelastic and Dielectric Effects in Polymeric Solids, Wiley, New York, 1967. Also see reference 4.
14. H. W. Starkweather, unpublished review.
15. C. W. Bunn and E. R. Howells, Nature 174, 549 (1954).
16. Spin-lattice relaxation times were measured and semi-quantitatively agree with those reported in reference 4. There is no indication that the α relaxation is detectable with T_1 measurements.
17. P. Diehl and C. L. Khetrapol, NMR, Basic Principles and Progress, 1, Springer-Verlag, New York, (1969).
18. N. O. Petersen and S. I. Chan, Biochemistry 16, 2657 (1977).
19. D. F. Bocian and S. I. Chan, Ann. Rev. Phys. Chem. 29, 307 (1978).
20. T. W. Bates and W. H. Stockmayer, Macromolecules 1, 12, 17 (1968).
21. R. G. Brown, J. Chem. Phys. 40, 2900 (1964).
22. G. M. Martin and R. K. Eby, J. Res. NBS, A 72A, 467 (1968).

23. E. S. Clark, J. Macromol. Sci.-Phys., B1(4), 795 (1967).
24. J. F. Rabolt, G. Piermarini, S. Block, J. Chem. Phys. 69, 2872 (1978).
25. a. J. F. Rabolt and B. Fanconi, Polymer 18, 1258 (1977).
    b. D. H. Reneker and B. Fanconi, J. Appl. Phys. 46, 4144 (1975).
26. D. H. Reneker, B. M. Fanconi, and J. Mazur, J. Appl. Phys. 48, 4032 (1977).
27. E. S. Clark and L. T. Muus, Int. Union of Crystallography, 6th Int. Congress, Rome, 1963, A-96; Polymer Preprints 5, 17 (1964).
28. P. G. de Gennes, J. Chem. Phys. 55, 572 (1971).
29. The numerical method used to decompose the chemical shift spectra are detailed in a paper submitted to Macromolecules.

RECEIVED June 10, 1980.

10

Variable-Temperature Magic-Angle Spinning Carbon-13 NMR Studies of Solid Polymers

W. W. FLEMING, C. A. FYFE[1], R. D. KENDRICK, J. R. LYERLA, H. VANNI, and C. S. YANNONI

IBM Research Laboratory, San Jose, CA 95193

It is now well-established ([1-5](#)) that ^{13}C NMR spectra in which resolution approaches that of the liquid state can be obtained for solid hydrocarbon polymers by employing a combination of resolution (dipolar decoupling, DD, and magic-angle spinning, MAS) and sensitivity (cross-polarization, CP) enhancement techniques ([6-8](#)). Although initial studies on solids by ^{13}C NMR have been carried out almost exclusively at ambient temperature, full exploitation of this spectroscopy requires variable temperature capability. This is particularly true of macromolecules since their physical properties are strongly influenced by temperature. Thus, the accessibility of variable temperature magic-angle spinning, VT-MAS, makes feasible the investigation of structural and motional features of polymers above and below T_g and in temperature regions of secondary relaxations. In this paper, we describe a spinner assembly suitable for routine operation over a wide range of temperature and report initial low temperature spectral data on several polymers, but with emphasis on fluoropolymers.

Background to Magic Angle Spinning for ^{13}C NMR in Solids

Preliminary to the description of the VT-MAS assembly, we describe briefly the function of MAS in the combined resolution/sensitivity enhancement experiment for solids.

---

[1] Mailing address: Guelph–Waterloo Centre for Graduate Work in Chemistry, University of Guelph, Guelph, Ontario, Canada N1G2W1.

Briefly, the chemical shift is a second-rank tensor which, because of rapid molecular tumbling, is averaged to its isotropic value, i.e., 1/3 the sum of the principal elements, in solution. In a polycrystalline or amorphous solid, the orientation dependence of the chemical shift is not averaged-out and as a consequence can result in broadening of a resonance line. In particular, unless the electronic shielding of a nucleus is spherically symmetric, the chemical shift of the nucleus will depend on the orientation of the principal axes of the shielding tensor with respect to the magnetic field direction. The NMR spectrum of a powder then consists of a superposition of lines encompassing the frequency range between the maximum and minimum values of the shielding - in essence, a powder pattern. The calculated lineshapes for an isotropic probability distribution (powder) of tensor orientations for uniaxially symmetric and non-axially symmetric tensors are presented in Figure 1 (6). The principal values of the shielding tensor, $\hat{\sigma}$, can be obtained directly from the spectrum as indicated. In Figure 1 are the respective carbon-13 spectra of solid benzene (at -40°C) and a highly crystalline poly(ethylene), PE, obtained under dipolar-decoupled conditions. At -40°C, the rapid reorientation of the benzene ring about the six-fold axis reduces the shielding tensor to one of axial symmetry. The non-axial symmetry of the shielding in PE is quite apparent. Knowledge of the principal elements of $\hat{\sigma}$ and of the effect of temperature (and thereby molecular motion) on the shape and width of the powder pattern provides a source of important information on a molecule. Benzene and PE (if chain ends are ignored) represent systems in which there is only one carbon resonance; in the more general case, where a molecule or polymer repeat unit possesses several magnetically-inequivalent carbons, the anisotropy in the chemical shielding will usually give rise to an NMR spectrum composed of overlapping powder patterns which make it virtually impossible to extract structural information in any straightforward manner. To obtain a more highly resolved spectrum requires the averaging of the various carbon shielding tensors in the solid state.

The required averaging can be induced by utilizing the magic-angle spinning technique introduced into NMR by Andrew (9) and Lowe (10) to attempt removal of dipolar interactions in solids. (The removal of C-H dipolar interactions is accomplished by proton-decoupling for the case of ^{13}C NMR in solids.) The expression for the chemical shift of a nucleus i with principal shielding elements σ_{i1}, σ_{i2}, σ_{i3} and with direction cosines between the principal axes of $\hat{\sigma}$ and $\vec{H}_0$, λ_{i1}, λ_{i2}, λ_{i3} is given by (9)

$$\sigma_{izz} = \lambda_{i1}^2 \sigma_{i1} + \lambda_{i2}^2 \sigma_{i2} + \lambda_{i3}^2 \sigma_{i3} \,. \qquad (1)$$

If the solid sample in which i is resident is rotated at angular velocity ω_r about an axis inclined at an angle β to $\vec{H}_0$ and at

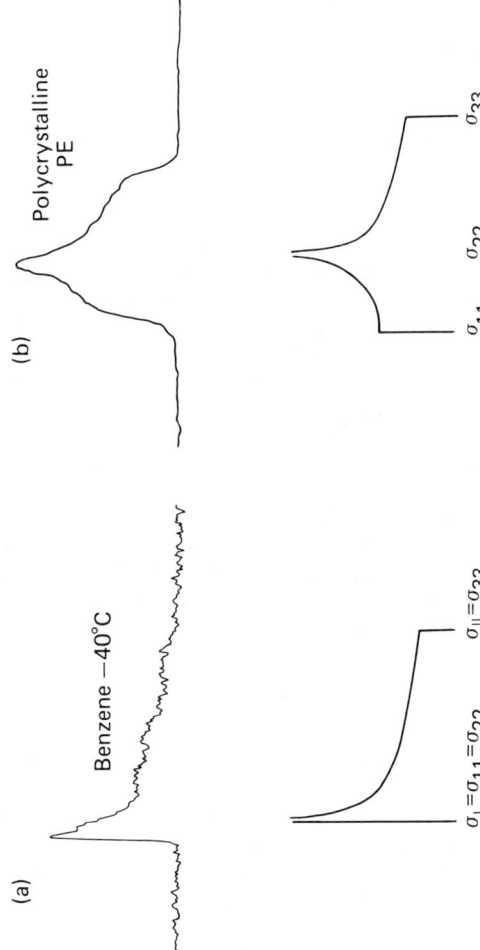

Figure 1. Calculated absorption lineshapes and representative C-13 spectra for polycrystalline samples: (a) an axially symmetric chemical shielding tensor ($\sigma_\parallel$ and $\sigma_\perp$ denote shielding parallel and perpendicular to the symmetry axis) e.g., the spectrum of frozen benzene at $-40°C$; (b) a general shielding tensor, e.g., the spectrum of polycrystalline PE (47)

angles ϕ_{i1}, ϕ_{i2}, ϕ_{i3} to the principal axes of $\hat{\sigma}$, the direction cosines become

$$\sum_{k=1}^{3} \lambda_{ik} = \cos\beta\cos\phi_{ik} + \sin\beta\sin\phi_{ik}\cos(\omega_r t + \chi_{ik}) , \qquad (2)$$

where χ_{ik} is the azimuthal angle of the kth principal axis at t=0. Substitution of (2) into (1) and taking the time-averaged yields

$$\bar{\sigma}_{izz} = 1/2\sin^2\beta(\sigma_{i1}+\sigma_{i2}+\sigma_{i3}) + 1/2(3\cos^2\beta-1)\sum_{k=1}^{3}\sigma_{ik}\cos^2\phi_{ik} . \qquad (3)$$

If β is the magic angle (54.7°), (3) reduces to

$$\bar{\sigma}_{izz} = 1/3(\sigma_{i1}+\sigma_{i2}+\sigma_{i3}) , \qquad (4)$$

which is the average value obtained in fluids. The result holds for all nuclei in the sample irrespective of the initial orientation of the principal axes of $\hat{\sigma}$ with respect to H_0 and to the spinning axis.

In the first approximation, in order to remove line broadening due to shielding anisotropy the spinning frequency should be fast compared to the anisotropy expressed in frequency units or equivalently to the frequency range encompassed by the powder pattern. Typical shift anisotropy values for aliphatic carbons are 15-50 ppm while for aromatic and carbonyl carbons values of 120-200 ppm are observed (6). These data reflect the larger directional variation in electron density (and hence shielding) associated with multiple bonds. At an applied field of 1.4T, the anisotropy for aromatic and carbonyl carbons translates to ca. 1.8-3 kHz. Thus, spinning speeds greater than 3 kHz would lead to removal of the broadening.

Spinning Apparatus

Two main techniques have been used to produce the high spinning rates required for MAS, one in which the rotors are supported by gas bearings and the other in which rotors are supported by an axle assembly. The latter design, due to Lowe (10), usually employs a cylindrical rotor which spins on a horizontal axle of phosphor-bronze. The rotor is driven by compressed gas directed from a jet onto peripheral vanes on the rotor's exterior. The gas bearing design, pioneered by Beams and adapted to NMR by Andrew, requires a stator/rotor assembly with the rotors having a conical underface and a cylindrical stack (see Figure 2). Compressed gas enters the stator through an inlet tube and is directed through inclined jets onto the rotor's flutes. The rotor is thus supported and driven by the gas. The advantages of the Lowe design are the ease of loading the sample cylinder into the rf coil of the probe, ease of reaching the magic angle, and ease of adapting the assembly to variable temperature operation (12). Disadvantages of the design include rapid wear of the rotor (a polymer) by the axle, sensitivity of spinning to rotor balance, and the small filling factor of the rf coil volume by the sample. The advantages of the gas bearing design are reduced wear, less sensitivity to rotor balance, and a good filling factor as the cylinder stack fits precisely into the coil (the cone of the rotor is outside the coil) which is canted at the magic angle rather than normal to the d.c. field. At this angle, the amplitude of the perpendicular component of H_1 is $0.816\ H_1$. This results in a voltage induced by transverse magnetization reduced by the same factor. The loss in S/N is compensated (under MAS conditions) by the better filling factor relative to a perpendicular arrangement.

The spinner assembly we have employed is a new design (due to Fyfe, Mossbruger and Yannoni (13)) whose characteristics represent something of a compromise of the two above-described techniques. The general philosophy behind the design is to provide for (1) ease of sample loading and operation using conventional designs of probes, and (2) for temperature variation.

The spinner assembly is shown schematically in Figure 2. The apparatus consists of four parts: holder, stator, rotor, and cover. The stator is threaded to the holder and the gas line (in the case of ambient temperature experiments) attached with a press fit. The rotor is inserted into the stator cup and the cover sleeve fitted over the stator. The general arrangement of stator cup and rotor is similar to that of Andrew (9). The gas flow enters the conical "donut shaped" cavity behind the stator cup (which is sealed to the outside by the sleeve), is directed toward the rotor through three small tangential holes (two of which are shown in the diagram) drilled through the stator cup, and impinges on the rotor. (The arrangement of rotor flutes and stator holes is such that a clockwise movement of the rotor is produced.) The exact shapes of these two components and the size and angles of inclination of the holes for the driving gas have to be carefully optimized as the apparatus makes a very tight fit into the probe insert and thus exit gas flow becomes important in stability considerations. The spinning axis is at right angles to the holder and stator assembly.

Spinner assemblies (i.e., stator, holder, rotor) have been constructed from three materials: Kel-F, Delrin, and machinable boron nitride (BN) (14). For observation of hydrocarbon materials at ambient and low temperatures, the Kel-F assembly is used. It displays suitable mechanical properties and does not interfere with the carbon spectrum since the resonances of the carbons in the Kel-F are >10 kHz in width due to the unremoved C-F dipolar interactions. To observe fluorocarbon materials at ambient and low temperature by C-F dipolar decoupling/CP/MAS, the Delrin assembly is used since unremoved C-H dipolar interactions broaden

the Delrin signal beyond detection. Finally, to obtain high
temperature spectra, the BN system, which is transparent in
^{13}C NMR, has been employed. Samples in the form of powders are
loaded into hollow rotors (volume ca. 65 μℓ) which are fitted with
threaded caps to confine the sample. No problems were encountered
with this type of arrangement provided the threads of the cap were
at or below the top rim of the rotor. In some cases, the polymer
under study was itself machined in the form of a rotor and used as
the analytical sample. The rotors have a cone diameter of 8 mm
and experience centripetal acceleration on the perimeter of
ca. 0.4 million G when rotating at 5 kHz (15).

For an electromagnet (as is employed on our spectrometer),
there are two alternative ways in which the magic-angle
arrangement can be achieved by continuous adjustment using the
spinner design. As illustrated in Figure 3 there is a cone of
possible 54°44' inclinations of the spinning axis to the magnetic
field vector. The correct angle can be obtained by inserting the
holder vertically into the probe (with the rotor spinning around a
horizontal axis) and rotating it until the correct angle is
reached, or by inserting the holder horizontally into the probe
(also mounted horizontally) with the rotor spinning around a
vertical axis and then rotating the holder until the correct angle
is reached. Although the first arrangement in reaching the angle
corresponds to a conventional rf-coil/H_0-field geometry, the
variable temperature hardware is more easily arranged with the
second geometry and, as such, is used on our spectrometer. In
both arrangements, the exact angle chosen is reached by continuous
adjustment of the rotation angle which is controlled by a
goniometer head attached to the top or front of the probe body
(see Figure 5). This gives considerable fine control over the
exact angle chosen in an experiment.

The results of various tests of the spinning performance of a
version of the apparatus designed for a 10 mm I.D. probe insert
and employing a solid rotor machined from Lexan (14) are shown in
Figure 4. (Qualitatively similar results are obtained for 15 mm
and 7.5 mm probe versions of the equipment with the efficiency of
the spinning found to increase markedly as the apparatus decreases
in size. The curves shown are the result of several experiments
using different rotors and stators and each curve represents
average values. Further, the curves obtained are found to be
mainly functions of the geometric arrangement in the design and
the propellant gas used. There is no significant variation found
in the spinning rate with the position of the spinning axis
(vertical or horizontal "magic angle" arrangement) or when rotors
packed with powered material are utilized. With N_2 as the
propellant gas and the system open to the atmosphere, Curve A is
obtained. There is little change in this curve when the system is
enclosed in a 10 mm I.D. glass tube. The maximum spinning rate of
1.5 kHz is adequate for experiments at ^{13}C resonance frequencies
up to 15 MHz.

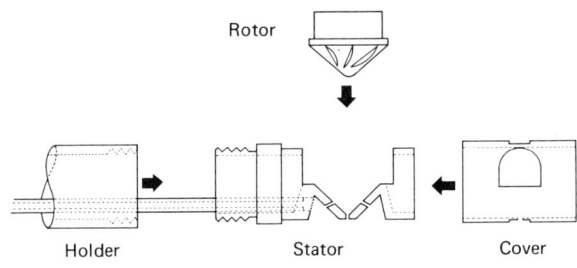

Figure 2. Schematic of the component parts of the spinning apparatus (side view) and their relationship to each other (see text) (13)

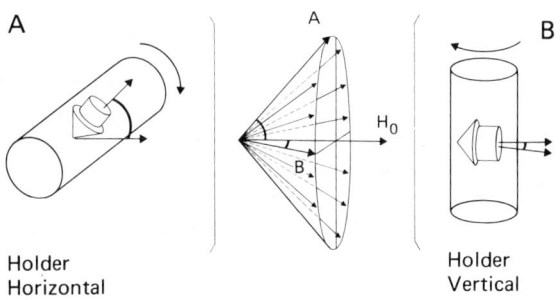

Figure 3. Schematic of the spinner device showing two possible ways in which the correct magic-angle setting of the inclination of the spinning axis to the magnetic field axis may be obtained by continuous adjustment (13)

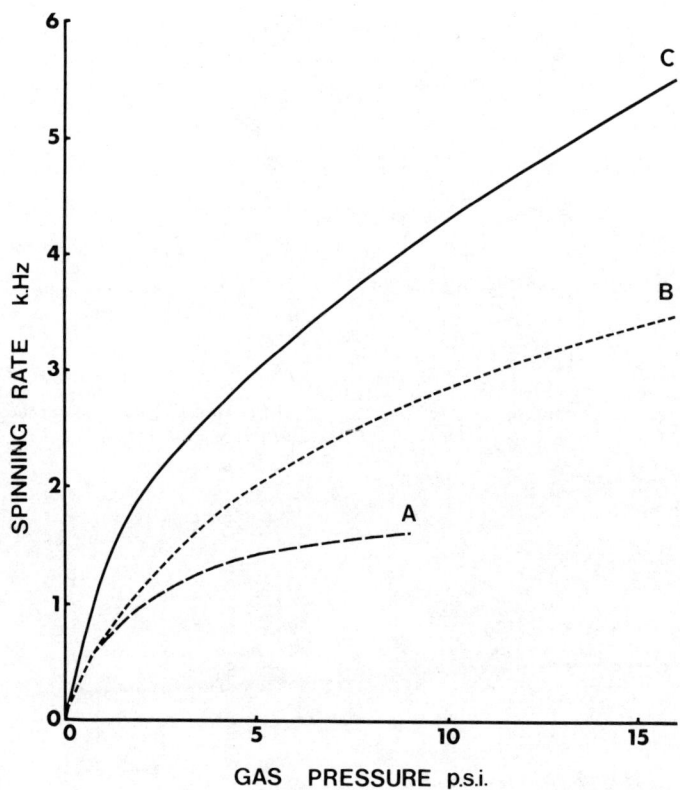

Journal of Magnetic Resonance

Figure 4. Plot of spinning rates (kHz) vs. applied gas pressures (in pressure per square inch) for a 10 mm version of the spinning apparatus. Curve A: sample with nitrogen gas with the apparatus in an open environment in air. Curve B: spun with helium gas with the apparatus in an open environment in the air. Curve C: spun with helium in a closed glass tube (11 mm I.D.) as in the probe arrangement. The sample was spun in an atmosphere of helium in this case. The three arrangements gave stable spinning over the range of pressures indicated in the figure and were followed for both solid and hollow rotors and for samples spun at different angular settings (13).

Since variable temperature operation down to at least liquid nitrogen temperature is desirable, the use of helium as propellant gas was investigated. Curve B shows the rates achieved for helium with the system open to the atmosphere. There is an increase in the rates over those for nitrogen gas, as would be expected from the much lower density of helium. There is, however, a further, very substantial increase in performance when the assembly is placed within a glass tube of inner diameter 10-11 mm as is the case in the actual probe insert (Curve C). This is thought to be due to a reduction in the friction between the rotor and the surrounding atmosphere of gas, which is now composed entirely of helium. At ambient temperature, and 15 psi pressure, spinning rates >5 kHz are readily achieved. (The maximum theoretical rate for an 8 mm diameter rotor with He as the propellant is ca. 38 kHz at 0°C. (16).) These rates are obtained on rotors without flutes. As fluting a rotor is a complicated machining operation, the ability to utilize smooth rotors is a desirable feature.

Temperatures below ambient are achieved by mixing He at ambient temperature with He precooled by flow through a heat exchanger immersed in liquid nitrogen. The mixture of gases is transferred by a Cu/Be Dewar (over which the spinning assembly fits) directly into the stator cup, resulting in spinning and cooling. Spinning rates up to 4 kHz are obtained even as low as 77°K with no problems of rotor stability being experienced. Temperatures above ambient are obtained by pre-heating the He at entry into the transfer Dewar. To the present, we have obtained spinning spectra over a range of -195°C to +100°C. Temperature is controlled to ±2°C with a homebuilt temperature sensing and heater/feedback network. The entire spinning assembly and variable-temperature accessories are mounted on an "optical rail" for straightforward and reproducible insertion of the spinner into the probe, which is fitted with a "straight-through" design of glass Dewar. The spinner/probe arrangement that is routinely used in our laboratory for variable-temperature studies is depicted in Figure 5. To prevent rf discharge of the He, the rf coil is isolated from the driving gas and sample holder by a glass tube fitted inside a Kel-F (14) coil-form (see Figure 5).

Spectrometer

The basic spectrometer we have employed for the study of solid polymers by cross-polarization/magic-angle spinning techniques is a Nicolet TT-14 pulse-FT system operating at 1.4T with resonance frequencies of 15.087 MHz, 60.0 MHz and 56.4 MHz for ^{13}C, ^{1}H and ^{19}F respectively. The magnet is a Varian HA-60 electromagnet with a 1 5/8" polegap. The spectrometer has been modified for DD/CP by: (1) constructing a probe capable of withstanding the rf power levels required; (2) adding external gating circuitry (for spin-locking, etc.) and rf amplifiers; and (3) rerouting some of the timing logic. The probe (10 mm I.D.) is a variation of the

double-tuned single coil design reported by Stoll, Vega and
Vaughan (17). The coil is tuned for ^{13}C resonance and either ^{1}H
or ^{19}F resonance. The probe has an external deuterium (D_2O) lock
system. Broadband (0.25-105 MHz) 100W rf power amplifiers
(ENI-3100L) are employed to amplify the basic output of the
spectrometer's ^{13}C and ^{1}H channels or the low level ^{19}F output
derived from a frequency synthesizer. With this arrangement, it
is possible to achieve a Hartmann-Hahn CP match condition up to
ca. 55 kHz for both the ^{13}C-$\{^1$H$\}$ experiment and the ^{13}C-$\{^{19}$F$\}$
experiment. During FID collection or if CP is not required, the
proton decoupling network is capable of sustaining an H_1 field of
25 gauss at 100W for dipolar decoupling, while the ^{19}F network
yields a 22 gauss H_1 field at 100W. To remove artifacts in the CP
spectrum which can arise from the long rf pulses in the observed
channel, data are collected using a scheme of alternating the
proton spin temperature (18).

Results and Discussion

In this section, initial spectral data on several polymers are
reported to illustrate the utility of variable-temperature
magic-angle spinning (VT-MAS) for the study of macromolecules in
the solid state.

Poly(Ethylene Oxide) - PEO. The importance of a variable
temperature capability is obvious if features such as barriers to
reorientational modes, effects of phase transitions on chain
mobility, the nature of secondary relaxations, etc. are to be
determined. However, VT-MAS also has utility purely from the
point of view of the efficiency of the cross-polarization process
usually required to obtain carbon spectra in solids. In part, the
efficiency of the rotating-frame double-resonance experiment
depends on the abundant spin (I) having a $T_{1\rho}^I$ of sufficient
duration to allow complete polarization transfer and T_1^I being
sufficiently short to allow rapid repolarization of the I-spins.
Since the two relaxation times are functions of temperature, a
variable temperature spinning capability allows for an optimally
efficient double resonance experiment to be performed. For
example, poly(ethylene oxide) has a minimum in the proton
rotating-frame relaxation time, $T_{1\rho}^H$, at ambient temperatures (20).
This results in a very weak ^{13}C signal using CP techniques since
proton order is destroyed before cross-polarization of the carbons
has proceeded to a significant degree. However, at temperatures
below ca. -60°C, $T_{1\rho}^H$ is sufficiently long to utilize CP methods
for sensitivity enhancement and thus make feasible study of PEO in
the bulk state. The spinning and non-spinning carbon spectra of
PEO (M_W=15,000) at -140°C are shown in Figure 6. The presence of
an upfield shoulder on the resonance in the MAS spectrum
(Figure 6a) may result from non-crystalline regions of the sample.
Such an observation has been reported in polyethylene (3,4) where

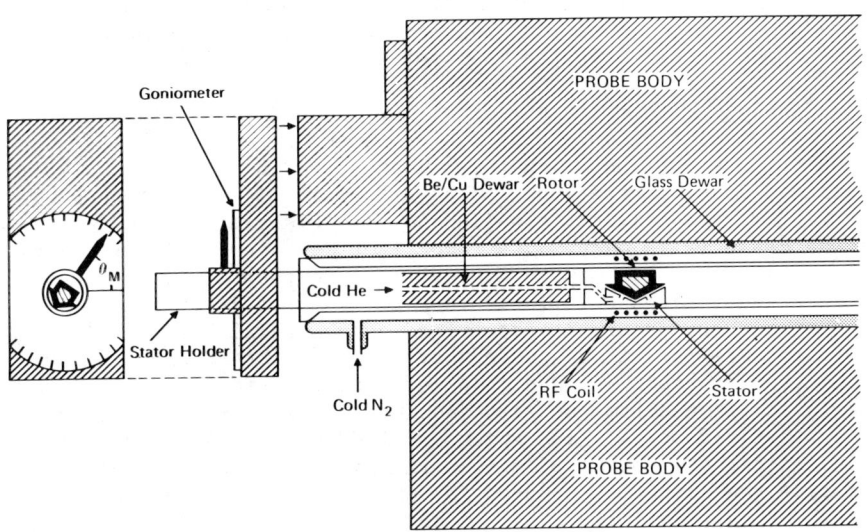

Figure 5. Schematic of the spinner assembly/probe geometry used for VT-MAS. The rf coil (silver ribbon) is wound on either a Kel-F or Delrin (14) form.

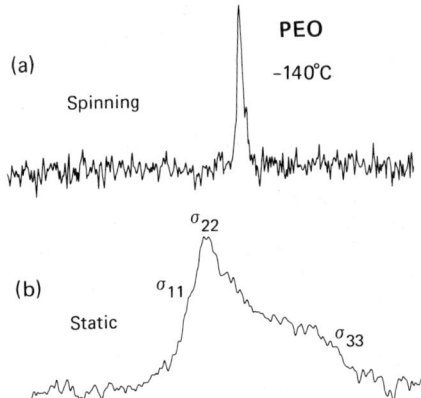

Figure 6. Proton-decoupled/CP C-13 NMR spectra of PEO at $-140°C$: (a) under MAS conditions; (b) nonspinning. The three principal elements of the shielding tensor are indicated in the static spectrum.

constraint to trans conformation in crystalline regions results in
a 2.3 ppm downfield shift relative to the resonance from
non-crystalline regions. The anisotropy pattern at -140°C in PEO
is non-axially symmetric and the values of the principal element
of the shielding tensor are 91±2, 83±2 and 33±2 ppm. These data
yield a value of 69 ppm for the isotropic shift as compared to
70.4 ppm determined in the MAS spectrum and solution.

Cis-1,4 Polybutadiene - CPBD. At ambient temperatures, CPBD
is elastomeric and molecular motion is sufficient to average C-H
dipolar interactions. Thus, a ^{13}C spectrum (Figure 7a) can be
obtained using conventional pulse - Fourier transform techniques.
At -150°C, the polymer is well-below its glass transition
temperature (-102°C) and the removal of C-H dipolar interactions
requires high power proton decoupling. In the resulting spectrum
(displayed in Figure 7c), the resonances of both vinyl and
methylene carbons are characterized by non-axially symmetric
powder patterns. As discussed earlier, the larger chemical shift
anisotropy for the vinyl carbon reflects the greater directional
variation in electron density (and hence shielding) associated
with multiple bonds. The spectrum of CPBD at -150°C under MAS
conditions (ca. -3 kHz) is given in Figure 7b. Both resonances,
despite the large differences in anisotropy, have the same full
width at half-maximum (FWHM) of ca. 60 Hz and sidebands are absent
from the spectrum, demonstrating the efficiently of the spinning
device at low temperature. The chemical shift values are found to
be within ±1 ppm of those found at ambient temperature

It is evident that in the static ^{13}C spectrum of CPBD at
-150°C, only two elements of the shielding tensor for the vinyl
carbon are resolved. The third overlaps the -CH_2- pattern, thus
making it impossible to fully assign the tensor. However,
variable angle spinning allows the full assignment to be made
through the angular functional dependence of the spectrum. The
spinning assembly described above makes possible spinning at the
full complement of rotational angles (0°-90°). Thus, ^{13}C spectra
can be obtained in which the chemical shift anisotropy is scaled
by $P_2(\cos\theta)$, where θ is the angle between the field and the
spinning axis (see Figure 7d). For example, with the sample
spinning at an angle of 87°, the $P_2(\cos\theta)$ functional dependence
predicts a spectrum, relative to the static case, reduced in width
by 0.492 and with the principal elements of the shielding tensor
reversed with respect to the isotropic shift (21). The principal
values of $\hat{\sigma}$ calculated from the data of Figure 7d are $\sigma_{11}=236\pm4$,
$\sigma_{22}=115\pm4$, $\sigma_{33}=35\pm4$. Assuming a similar orientation of the
shielding tensor as in ethylene ($\sigma_{11}=236$; $\sigma_{22}=124$; $\sigma_{33}=27$ (22)),
these elements correspond to shielding approximately perpendicular
(but in plane) to the double bond direction, parallel to the bond
direction, and perpendicular to the plane of the bond
respectively. The isotropic shift obtained from these values is

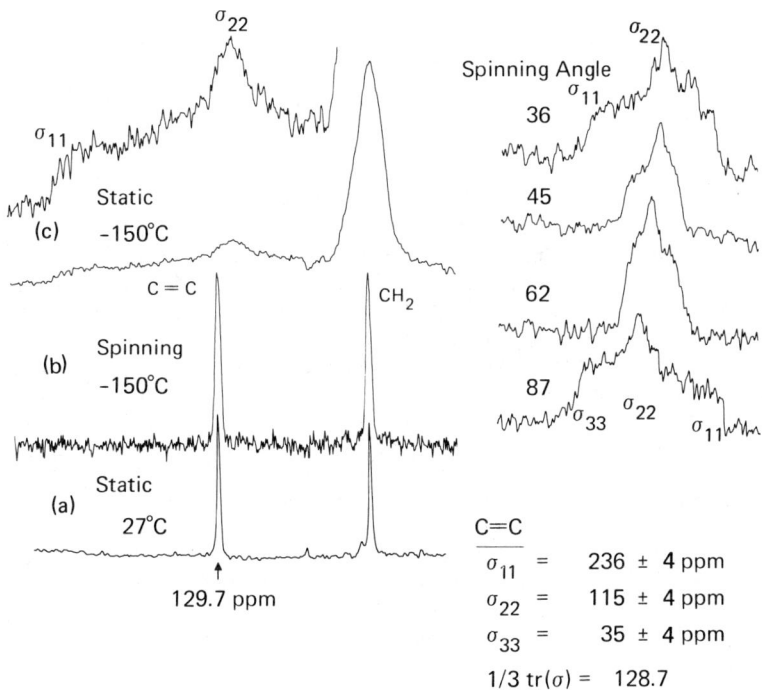

Figure 7. C-13 NMR spectra of cis-1,4 polybutadiene: (a) ambient temperature, $\pi/2$ pulses, ca. 1 G decoupling field; (b) CP/MAS at $-150°C$, 22 G decoupling field; (c) CP/decoupled, nonspinning at $-150°C$ — two elements of the shielding tensor are indicated for the vinyl carbon in the ×4 vertical expansion (inset); (d) CP/decoupled spectra as a function of spinning angle. Tensor elements are indicated at the bottom (47)

128.7 ppm as compared to 129.7 determined from the spectrum in the elastomeric state and the MAS spectrum at -150.

In addition to allowing elements of the shielding tensor to be assigned, the variable-angle VT-MAS experiment offers the potential to observe, in polymers with multiple carbon resonances, the manner in which the shielding tensors undergo averaging as a function of temperature. Such data may provide insight into the nature of motions (i.e., along chain axis, perpendicular to chain axis, etc.) that enter a polymer chain as a function of temperature (and thereby physical state (23)).

Fluoropolymers. Fluoropolymers are subject to NMR analysis in the solid state by using the ^{19}F spins to cross-polarize the ^{13}C spins and employing ^{19}F decoupling and MAS for resolution enhancement. In principal, the narrowing of ^{13}C resonance lines in fluoropolymers by "solid-state" methods (6) is no more difficult than in hydrocarbons. However, there is a significant experimental difference between the two cases with respect to removal of the heteronuclear (C-H or C-F) dipolar broadening. In particular, consider that there exists a spread in the ^{1}H or ^{19}F resonance frequency resulting from non-equivalent nuclei in the sample and/or from chemical shift anisotropy. Effective dipolar decoupling requires that the decoupling field be of sufficient magnitude that all spins causing the dipolar broadening experience an effective field in the rotating frame that is close to being "on-resonance." (For details, the reader is referred to the book by Mehring (24).) Because the proton has a small chemical shift range and also small shielding anisotropies, decoupling in hydrocarbon polymers is effective with fields of ~40 kHz. On the other hand, the fluorine nucleus exhibits a range of isotropic chemical shifts ca. twenty-fold greater than that of the proton (25). In solids, typical fluorine chemical shift anisotropies display a corresponding relationship to their proton counterparts. (26). The consequence of such large dispersions in the fluorine resonance is that larger decoupling fields are required for removal of C-F dipolar broadening of carbon resonance lines than for C-H broadening in analogous cases. To illustrate, magic-angle spinning ^{13}C spectra of poly(tetrafluoroethylene), PTFE, (60% crystalline) obtained at -72°C as a function of the ^{19}F decoupling field strength (ω_{1F}) are presented in Figure 8 (27). For semicrystalline polyethylene, PE, spectra with FWHM of $\lesssim$5 Hz can be obtained from the crystalline region with 35-40 kHz of proton $\lesssim$40 Hz decoupling (4). At this same ^{19}F decoupling field, PTFE yields a 50-70 Hz resonance line having large wings at its base. The difference in resolution achieved in the two polymers is attributed mainly to less efficient dipolar decoupling in PTFE caused by "off-resonance" effects arising from the large ^{19}F chemical shift anisotropy (23,28,29). At the magnetic field of 1.4T, the non-axially symmetric ^{19}F powder pattern encompasses a frequency range of ca. 8 kHz as opposed to ~270 Hz frequency range for the ^{1}H powder pattern in PE (30). As is evident in Figure 8, the ^{13}C resonance line in PTFE does narrow significantly at large values of ω_{1F} with the minimum ^{13}C linewidth (15 Hz FWHM) being observed at the maximum available decoupling field (89 kHz). The smooth decrease in PTFE linewidth up to the maximum decoupling field (31) suggests that residual fluorine dipolar broadening accounts, in part, for the 15 Hz linewidth observed at this field strength.

^{13}C magic-angle spinning spectra of PTFE as a function of temperature are shown in Figure 9. The linewidth (15 Hz) of the CF_2 resonance remains essentially constant until the temperature regime (15°-25°C) is reached above which a broadening of 2-3 fold (to 35-45 Hz) ensues. The broadening appears to reflect the 19°C first order transition which has been observed in PTFE by numerous techniques including ^{19}F broad-line NMR (32) and more recently by ^{19}F NMR using a multi-pulse narrowing scheme (23,29). The transition is associated with the onset of rotation of molecules in crystalline regions about the long axis of the helical chain (32). The resulting anisotropic motion is sufficient to give rise to motional broadening (1) of the carbon resonance. The same temperature-induced broadening phenomenon is also observed for a 95% crystalline sample of PTFE, thus establishing that the effect occurs in crystalline regions.

Magic-angle spinning ^{13}C spectra of poly(chlorotrifluoroethylene) PCTFE obtained as a function of temperature are shown in Figure 10. At 48°C, the two carbon resonances of the repeat unit are clearly resolved with the line at higher field being assigned to the CFCℓ carbon. Linewidths in the spectrum are ca. 60-70 Hz FWHM. Although the experimental conditions used to obtain the spectrum utilized a ^{19}F decoupling field only equal to the Hartmann-Hahn (8) matching field of 57 kHz, at higher values of ω_{1F} the widths of the lines are not changed significantly nor are they greatly affected by variation of the frequency of ω_{1F} over several kHz. This may indicate that there is considerable broadening of the resonance lines due to a distribution of isotropic chemical shifts arising from the stereochemistry of the polymer, as has been observed for several of the carbons in glassy PMMA (1).

As the temperature is lowered, the CFCℓ resonance begins to broaden and the appearance of a two-line spectrum is lost. The source of broadening is not C-F dipolar coupling as the ^{19}F rigid lattice linewidth is reached at ca. 20°C (33) and the FWHM of the CF_2 carbon resonance does not change measurably above or for a 100°C range below this temperature. Thus, the differential broadening of the CFCℓ resonance is attributed to a residual dipolar interaction of this carbon with the quadrupolar chlorine nucleus. In organic systems, chlorine nuclei often experience large quadrupole coupling constants (34). Depending on the ratio of the Zeeman and quadrupole interaction energies, the chlorine spins may be oriented along the Zeeman field or fixed at some

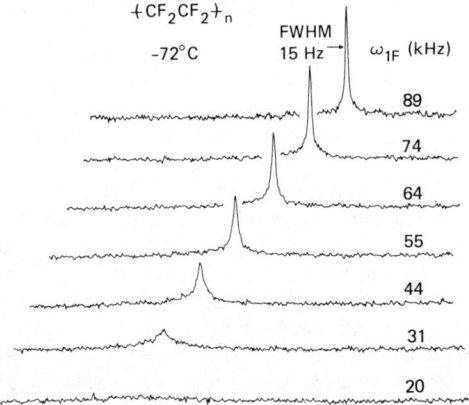

Figure 8. Magic-angle spinning C-13 spectra of PTFE at $-72°C$ as a function of the ^{19}F decoupling field, ω_{1F}, in kiloHertz. Each spectrum was obtained from a 2K FT of 256 FID accumulations with a CP time of 3.0 ms and an experiment repetition time of 2 s. The spectra are displayed on a normalized scale. For further details, see Ref. 27.

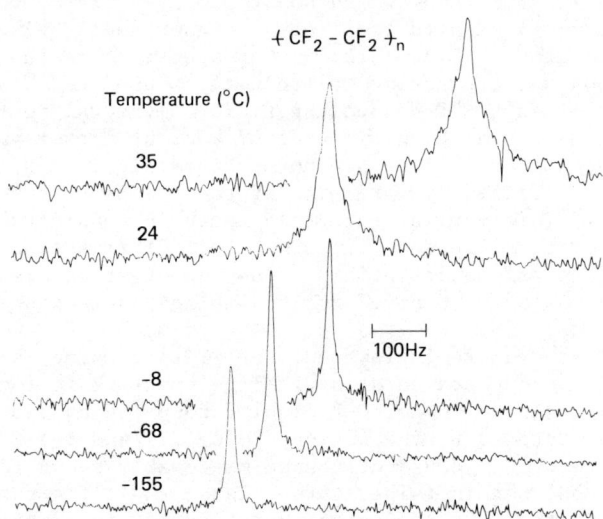

Figure 9. Fluorine-decoupled, MAS C-13 spectra of PTFE as a function of temperature. Spectra were obtained from a 2K FT of 300 FID accumulations with a CP time of 5.0 ms and an experiment repetition time of 2.0s. The ^{19}F decoupling field was 89 kHz in each case.

definite orientation in the principal axis system of the molecular electric field gradient tensor (35). If the Cℓ moments are coupled more strongly to the electric field gradient than to the Zeeman field, the C-Cℓ dipolar interaction will not be removed by magic-angle spinning (36). In the Zeeman field of 1.4T, this interaction appears to broaden the C-Cℓ resonance in PCTFE at temperatures below ambient. The separation of the resonances at higher temperature appears to be coupled to molecular motion in the polymer. Glass transition temperatures (33) for PCTFE are ca. 40°. Thus, at 48°C, motion in the polymer backbone is presumably sufficient to cause the Cℓ spin-lattice relaxation rate to be rapid enough to induce near-complete "self-decoupling" of the C-Cℓ dipolar interaction (37). Well below T_g, the small amplitude of oscillatory motions may be insufficient to cause rapid quadrupolar relaxation and thus C-Cℓ dipolar broadening ensues. Effects on the CF_2 resonance are strongly attenuated by the r^{-3} dependence of the dipolar interaction; however, at temperatures below -135°C, the broadening of the CF_2 carbon is apparent.

In closing this section on fluoropolymers, we report the spectrum of semi-crystalline poly(vinylidene fluoride), PVF_2 (38). To obtain HR-NMR carbon spectra of solid PVF_2 requires the simultaneous elimination of dipolar broadening from both proton and fluorine nuclei. This is accomplished using a probe with a single rf coil tuned to $^{13}C/^{1}H$ resonance and a surrounding "saddle"-coil tuned to ^{19}F resonance. Carbon magnetization is created by a C-H Hartmann-Hahn cross-polarization contact, then the ^{13}C FID is observed under conditions of strong ^{1}H and ^{19}F irradiation. The spectrum obtained from the triple resonance experiment (39) on a non-spinning sample is shown in Figure 11. The powder patterns yield principal elements for the shielding tensors (relative to TMS) of ca. 131, 120, 111 ppm for CF_2 and 56, 49, 30 ppm for CH_2. As compared to the shielding tensor of polyethylene, the major perturbation of the fluorines on the carbon tensor is a strong deshielding along the chain direction of an all-trans chain. Magic-angle spinning of a powder sample loaded in a boron-nitride rotor yielded a narrowed spectrum; however, the background resonance produced by either Delrin or Kel-F spinner assemblies (necessarily obtained under conditions of both ^{1}H and ^{19}F irradiation) has hampered further studies on this interesting compound.

Poly(Propylene). In principle, resolution of individual carbon resonances in bulk polymers allows relaxation experiments to be performed which can be interpreted in terms of mainchain and side chain motions in the solid. This is a distinct advantage over the more common proton NMR relaxation experiments where efficient spin-diffusion usually results in the averaging of the relaxation behavior over the ensemble of protons. Thus, a direct

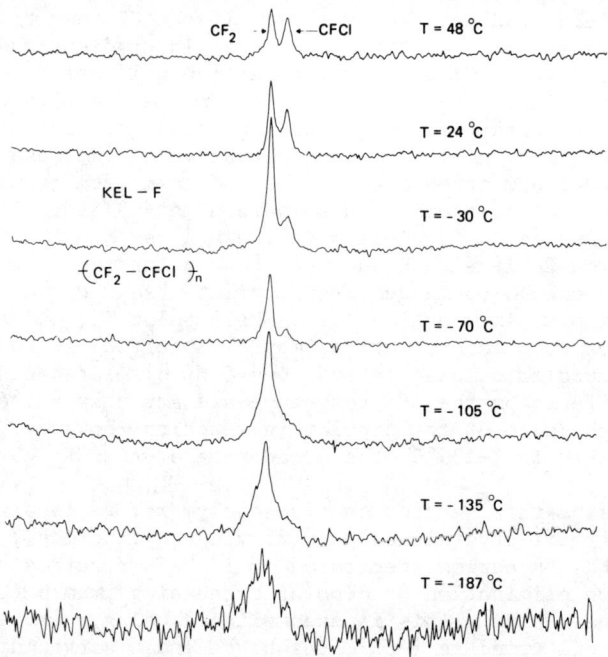

Figure 10. Fluorine-decoupled, MAS C-13 spectra of PCTFE as a function of temperature. Spectra were obtained with a CP time of 5.0 ms and an experiment repetition time of 2.5 s. Spectra represent a 2K FT in each case; the number of FID accumulations varied from 1K to 5K (47).

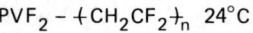

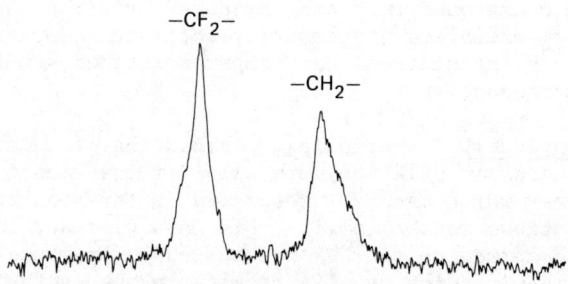

Figure 11. The C-13 NMR spectrum of semicrystalline PVF_2 obtained with C—H CP and simultaneous decoupling of the C—H, C—F dipolar interactions

interpretation of ^1H relaxation data in terms of unique motions of the solid polymer is often not possible.

Relaxation parameters of interest for the study of polymers include (1,2,5,40): the ^{13}C spin-lattice relaxation time in H_0, T_1^C; the spin-spin relaxation time, T_2; the nuclear Overhauser enhancement, NOE, the proton and carbon rotating frame relaxation times, $T_{1\rho}^x$; the C-H cross-polarization or cross-relaxation time, T_{CR}; and the proton relaxation time in the dipolar field, T_{1D}. Not all of these parameters provide information in a direct manner; nonetheless, the inferred information is important in characterizing motional frequencies and amplitudes in macromolecules. Fundamental to this characterization is the measurement of relaxation data over a range of temperature. As an example, we report in this section very preliminary results on the T_1 and $T_{1\rho}$ behavior for the carbons in a 90% isotactic, 70% crystalline sample of poly(propylene), PP.

The proton-decoupled, CP/MAS ^{13}C spectra of PP at 24°C and -170°C are shown in Figure 12. At ambient temperature, all three carbons of the PP repeat unit are resolved. The methylene and methine carbons are separated by 17.8±0.5 ppm while the methyl and methine resonances are separated by 4.9±0.5 ppm. These values are compared to separations of 17.6 and 7.0 ppm observed for this PP in 1,2,4 trichlorobenzene at 110°C. A similar downfield shift of the methyl resonance in the solid state has been reported very recently by Balimann, et al. (41). The shift may arise from conformational differences between the solid and solution; in particular, a reduction in the number of methyl carbon gauche interactions in the crystalline state could account for the downfield shift (42). This speculation is currently under investigation. When the temperature is lowered to -170°C, the methyl resonance is lost in the baseline. The broadening may arise, in part, from incomplete decoupling as the reorientation rate of the methyl group about the C_3-axis becomes comparable to the strength of the decoupling field. Heteronuclear dipolar couplings characterized by correlation frequencies near the decoupling frequency are not efficiently decoupled. This phenomenon has also been observed by Garroway for methyl groups in epoxy resins (43). As indicated, the results on PP are only preliminary and thus other explanations for broadening based on a distribution of isotropic chemical shifts merit consideration.

The initial relaxation results on PP over a temperature range from 24°C to -195°C are summarized in Figure 13. The T_1 data were collected using a pulse sequence developed by Torchia (44) which allows cross-polarization enhancement of the signals. The $T_{1\rho}$ data were determined at 58 kHz using $T_{1\rho}$ methodology of Schaefer et al. (1). As indicated in the figure, each of the carbons display individual relaxation rates in both types of experiments. The CH and CH_2 carbons have a T_1 minimum at ca. -110°C, nearly the same temperature as that reported by McBrierty et al.(45) for the proton T_1 minimum in isotactic PP.

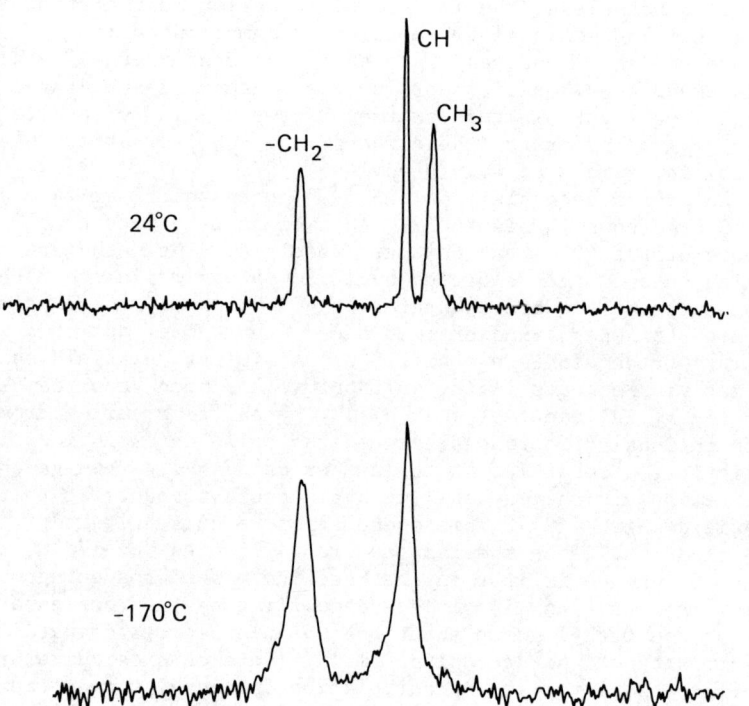

Figure 12. Proton-decoupled CP/MAS C-13 spectra of isotactic PP at 24°C and −170°C

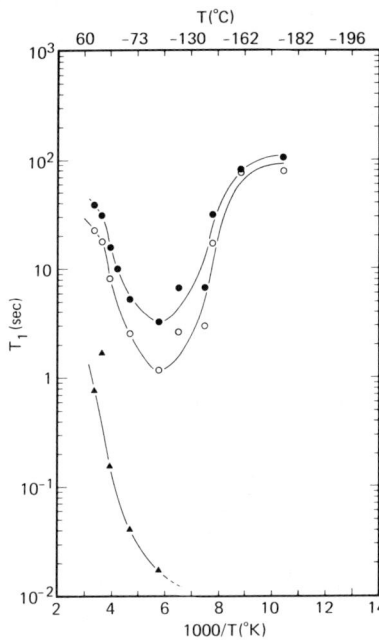

Figure 13A. C-13 relaxation times for the methyl (▲), methylene (●), and methine (○) carbons of PP as a function of temperature: spin-lattice relaxation times at 1.4T

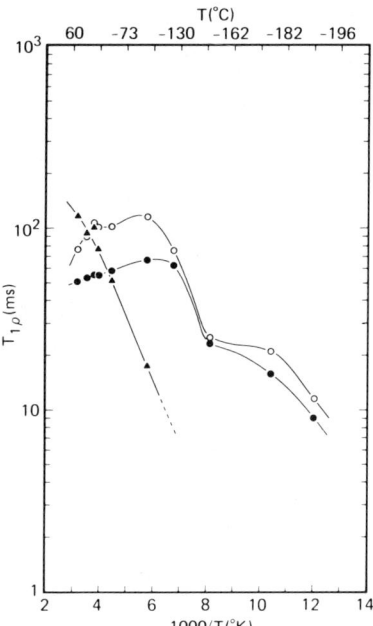

Figure 13B. C-13 relaxation times for the methyl (▲), methylene (●), and methine (○) carbons of PP as a function of temperature: rotating-frame relaxation times ($T_{1\rho}$) at 58 kHz

This result suggests, if it is assumed that a C-H heteronuclear dipolar relaxation mechanism is operative, that methyl protons dominate the relaxation behavior of these carbons over much of the temperature range studied despite the $1/r^6$ dependence of the mechanism. The shorter T_1 for the CH as compared to the CH_2 then arises from the shorter C-H distances. Apparently, the contributions to spectral density in the MHz region of the frequency spectrum due to backbone motions is minor relative to the sidegroup motion. The $T_{1\rho}$ data for the CH and CH_2 carbons also give an indication of methyl group rotational frequencies. As the temperature is lowered below -110°C, the contribution of the methyl protons to MHz spectral density decreases but increases in the kHz regime. Consequently, the $T_{1\rho}$ decreases by roughly 10× between -100°C and -195°C.

The interpretation of carbon $T_{1\rho}$ data is complicated by the fact that spin-spin (cross-relaxation) processes as well as rotating frame spin-lattice processes may contribute to the relaxation (40). Only the latter process provides direct information on molecular motion. For the CH and CH_2 carbons of PP, the $T_{1\rho}$'s do not change greatly over the temperature interval -110°C to ambient and, as opposed to the T_1 behavior, the CH_2 carbon has a shorter $T_{1\rho}$ than the CH carbon. These results suggest that spin-spin processes dominate the $T_{1\rho}$ (46). However, below ca. -115°C, the $T_{1\rho}$'s for both carbons shorten and tend toward equality. McBrierty et al. (45) report a proton T_1 minimum (which reflects methyl group reorientation at KHz frequencies) at -180°C. No clear minimum is observed in the ^{13}C data, perhaps due to an interplay of spin-spin and spin-lattice processes. Nonetheless, it is apparent that the methyl protons are responsible for the spin-lattice portion of the $T_{1\rho}$ relaxation for CH and CH_2 carbons.

Relaxation data for the methyl carbon could be measured only down to ca. -125°C; below this temperature line broadening was too severe to obtain results. The near equal values observed for the T_1 and $T_{1\rho}$ over much of the limited temperature interval is in accord with the methyl motion being on the high temperature side of the T_1-minimum.

As indicated earlier, these relaxation results are only preliminary; more extensive measurements are in progress aimed at fully explaining the relaxation behavior and deriving information on the polymer motions, including activation energies.

Acknowledgments

The authors would like to acknowledge the advice of Dr. D. Torchia (National Institute of Health) who made the details of his probe design and spectrometer modification available to us. The authors would also like to acknowledge the expert advice of Don Horne in many of the electronic aspects of the spectrometer modification.

Abstract

From the viewpoint of polymer applications, the full exploitation of the combined resolution/sensitivity enhancement techniques to obtain "high-resolution" spectra of rare-spin nuclei in solids requires variable temperature spinning capability. In this paper, we describe briefly a spinner assembly suitable for routine operation over a wide range of temperature at the full complement of spinning angles and report ^{13}C spectral data at low temperature on several polymers, including fluoropolymers. In addition, variable temperature ^{13}C spin-lattice and rotating frame relaxation times are reported for isotactic poly(propylene).

References and Notes

1. Schaefer, J.; Stejskal, E. O.; Buchdahl, R., Macromolecules, 1977, 10, 384,
2. Lyerla, J. R., "Contemporary Topics in Polymer Science," Vol. 3, p. 143, Shen, M., editor, Plenum Publishing Corp., New York, New York, 1979.
3. Fyfe, C. A.; Lyerla, J. R.; Volksen, W.; Yannoni, C. S., Macromolecules, 1979, 12, 757.
4. Earl, W.; VanderHart, D. L., Macromolecules, 1979, 12, 762.
5. Garroway, A. N.; Moniz, W. B.; Resing, H. A., ACS Symposium Series, 1979, 103, 67.
6. Pines, A; Gibby, M. G.; Waugh, J. S., J. Chem. Phys., 1973, 59, 569.
7. Andrew, E. R., MTP International Review of Science, Phys. Chem. Series Two, 1976, 4, 173.
8. Hartmann, S. R.; Hahn, E. L., Phys. Rev., 1962, 128, 2042.
9. Andrew, E. R., in "Progress in Nuclear Magnetic Resonance Spectroscopy," Emsley, J. W.; Feeney, J.; Sutcliffe, L. H., editors, Pergamon Press, New York, New York, 1972, Vol. 8, Ch. 1.
10. Lowe, I. J., Phys. Rev. Letters, 1959, 2, 285.
11. Beams, J. W., J. Appl. Phys., 1937, 8, 795.
12. Variable temperature studies of epoxy resins using a spinning device based on the Lowe design have been reported in Reference 5.
13. Fyfe, C. A.; Mossbruger, H; Yannoni, C. S., J. Magn. Resonance, 1979, 36, 61.
14. Teflon and Delrin are registered trademarks of the E. I. duPont Nemours and Co. (Inc.); Lexan is a registered trademark of the General Electric Co.; Kel-F is a registered trademark of the 3M Co.
15. Despite numerous start-ups, the Kel-F and Delrin rotors exhibit little wear after several hundred hours of operation over a wide range of temperature.
16. Determined from the maximum linear velocity of sound in He and the diameter of the rotor periphery ($\omega=(v^2/r^2)^{1/2}$).

17. Stoll, M. E.; Vega, A. J.; Vaughan, R. W.; Rev. Sci. Instrum., 1977, 48, 800.
18. Stejskal, E. O.; Schaefer, J., J. Magn. Resonance, 1975, 18, 560.
19. The desired ratio of $T_{1\rho}^I/T_1^I$ being as close to unity as possible. Of course, account must be taken of the other factors which affect the recycle time such as the time required to collect data for the desired resolution (i.e., digital resolution), etc.
20. Connor, T. M., in "NMR Basic Principles and Progress," Vol. 4, Springer-Verlag, New York, New York, 1971, pp. 247-269.
21. Mehring, M., "High Resolution NMR Spectroscopy in Solids," Springer-Verlag, New York, 1976, pp. 170-174.
22. Zilm, K. W.; Alderman, D. W.; Grant, D. M., J. Magnetic Resonance, 1978, 30, 563.
23. English, A. D.; Vega, A. J., Macromolecules, 1979, 12, 353.
24. Reference 21, Ch. 4.4.
25. Bovey, F. A., "Nuclear Magnetic Resonance Spectroscopy," Academic Press, New York, 1969, pp. 211-214.
26. Reference 21, pp. 153-166.
27. At this temperature both the crystalline and amorphous regions have ^{19}F linewidths close to the rigid-lattice value (see Reference 32). The spectra were obtained using a single contact, spin-locked, cross polarization experiment to produce carbon polarization. For each spectrum, the Hartmann-Hahn condition for $^{19}F/^{13}C$ was matched at 57 kHz then the appropriate ^{19}F decoupling field was gated-on during observation of the free induction decay. The value of 89 kHz (22 gauss) for ω_{1F} represents the maximum field available with our present experimental conditions.
28. Mehring, M.; Griffin, R. G.; Waugh, J. S., J. Chem. Phys., 1971, 55, 746.
29. Garroway, A. N.; Stalker, D. C.; Mansfield, P., Polymer, 1975, 16, 171.
30. Burum, D. P.; Rhim, W. K., J. Chem. Phys., 1979, 71, 944.
31. At 89 kHz, the ^{19}F field is considerably greater than static C-F dipolar interactions and F-F dipolar interactions.
32. McCall, D. W.; Douglass, D. C.; Falcone, D. R., J. Phys. Chem., 1967, 71, 998, and references therein.
33. McBrierty, V. J., Polymer, 1974, 15, 503.
34. Dailey, B. P., J. Chem. Phys., 1960, 33, 1641.
35. Abragam, A., "The Principles of Nuclear Magnetism," Oxford University Press, London, 1961, Ch. VII.
36. Lippmaa, E.; Alla, M.; Kundla, E., presented at the 18th Experimental NMR Conference, Asilomar, California, April 1977.
37. Spiess, H. W.; Haeberlen, U.; Zimmerman, H., J. Magn. Resonance, 1977, 25, 55.
38. The composition of the sample was approximately 50% crystalline and primarily of the alpha form.

39. The spectrum was obtained with the proton and fluorine decoupling frequencies set on resonance for the respective ^{1}H and ^{19}F nuclei in PE and PTFE. The reasonably well-defined powder patterns imply effective decoupling and suggest the absence of any large Bloch-Siegert shifts, (e.g., see A. Abragam "The Principles of Nuclear Magnetism," Oxford Univ. Press, London 1961, pp. 567-568).
40. VanderHart, D. L.; Garroway, A. N.; J. Chem. Phys., 1979, 71, 2773.
41. Balimann, G.; Burgess, M. J. S.; Harris, R. K.; Oliver, A. G.; Packer, K. J.; Say, B. J.; Tanner, S. F.; Blackwell, R. W.; Brown, L. W.; Bunn, A.; Cudby, M. E. A.; Eldridge, J. W., J. Chem. Phys., 1980, 46, 469.
42. Tonelli, A. E., Macromolecules, 1978, 11, 565.
43. Garroway, A. N., Naval Research Labs, Washington, D.C., private communication.
44. Torchia, D. A., J. Magn. Resonance, 1978, 30, 613.
45. McBrierty, V. J., Douglass, D. C., Falcone, D. R., JCS Faraday Trans. II, 1972, 68, 1051.
46. If the cross-relaxation to the dipolar reservoir is longer than the proton spin-lattice relaxation, T_{1D}, then the $T_{1\rho}$ for the CH carbon will be ca. 1.7-2.0x that of the CH_2 carbon for spin-spin domination of the rotating-frame relaxation. This is roughly the result observed in the data displayed in Figure 13. The explanation is based on the approximate two-fold difference in the CH second moments for the two types of carbons (for full details see Reference 40).
47. Lyerla, J. R., Methods of Experimental Physics, 1980, 16A.

RECEIVED July 14, 1980.

11

Drug–DNA Interactions in Solution: Acridine Mutagen, Anthracycline Antitumor, and Peptide Antibiotic Complexes

DINSHAW J. PATEL

Bell Laboratories, Murray Hill, NJ 07974

The successful application of high resolution nuclear magnetic resonance (NMR) spectroscopy to monitor the structure and dynamics of the helix-coil transition of oligonucleotide duplexes (1-4) and transfer RNA (5-9) in solution have prompted efforts in our laboratory to extend these investigations to the polynucleotide duplex level in solution (10,11).

Proton and phosphorus NMR spectra of DNA in solution show very large line widths and poor spectral resolution despite recent demonstrations of time-averaged flexibility of the double-helical state (12-15). We have circumvented the sequence dependent dispersion of the chemical shifts by investigating synthetic DNA's of defined repeating nucleotide sequence.[1]

Thermodynamic and kinetic studies have demonstrated that the alternating purine-pyrimidine polynucleotide poly(dA-dT) folds into smaller duplexes which melt independently of each other in solution (16,17). The NMR spectrum should be considerably simplified since all base pairs are structurally equivalent due to the symmetry of the alternating purine-pyrimidine duplex. Further, the NMR resonances may exhibit moderate line widths in the duplex state due to segmental mobility resulting from the rapid migration of the branched duplexes along the polynucleotide backbone. Finally, since the smaller duplexes melt independently of each other, the duplex dissociation rate constants may be on the NMR time scale, so that the resonances shift as average peaks through the melting transition. These features suggest that well resolved base and sugar resonances may be observable for high molecular weight alternating purine-pyrimidine polynucleotides in the duplex state and this has been confirmed experimentally (18) in studies of the duplex to strand transition of poly(dA-dT) (mol. wt. $\sim 10^6$, ~ 2000 base pairs) as a function of temperature.[2]

The NMR resonances of the polynucleotide duplexes at the Watson-Crick protons, the base and sugar protons and the backbone phosphates would provide sufficient markers to monitor ligand-nucleic acid interactions in solution. Such an approach has great potential in adding to our current knowledge of the interactions

of antibiotics and carcinogens with nucleic acids in solution
(19-21) and provides experimental probes to investigate the role
of symmetry in such recognition processes (22,23).

We first describe the NMR parameters for the duplex to strand
transition of the synthetic DNA poly(dA-dT) (18) with occasional
reference to poly(dA-dU) (24) and poly(dA-5brdU) and the corresponding synthetic RNA poly(A-U) (24). This is followed by a comparison of the NMR parameters of the synthetic DNA in the presence
of 1 M Na ion and 1 M tetramethylammonium ion in an attempt to
investigate the effect of counterion on the conformation and
stability of DNA. We next outline structural and dynamical aspects of the complexes of poly(dA-dT) with the mutagen proflavine
(25) and the anti-tumor agent daunomycin (26) which intercalate
between base pairs and the peptide antibiotic netropsin (27) which
binds in the groove of DNA.

Synthetic DNA. Alternating Adenosine-Uridine Sequences

This contribution complements an earlier review (11) which
summarized our NMR research on synthetic DNA's and RNA's with
alternating inosine-cytidine and guanosine-cytidine polynucleotides and the structure and dynamics of ethidium-nucleic acid
complexes.

Hydrogen Bonding in Duplex State: The base paired duplex
state in nucleic acids can be readily characterized by monitoring
the exchangeable imino protons in H_2O solution (28-32). Earlier
studies have demonstrated that the thymidine H-3 imino proton of
nonterminal base pairs in nucleic acid duplexes are in slow exchange with solvent H_2O and resonate between 12 and 15 ppm. The
360 MHz proton NMR spectra of poly(dA-dT) in 0.1 M phosphate,
H_2O solution (6 to 14 ppm) were recorded between 0° and 55°C and
the exchangeable protons identified by comparison with the corresponding spectra recorded in ^2H_2O solution. The thymidine H-3
imino hydrogen-bonded resonance is observed at 13.0 ppm in the
spectrum of poly(dA-dT) at 25.5°C (Figure 1A) and its chemical
shift and line width dependence are plotted as a function of temperature in Figures 1B and 1C respectively.

The linear upfield shift of the exchangeable resonance with
increasing temperature (Figure 1B) reflects, in part, contributions from a temperature dependent premelting conformational
change which is also observed for the nonexchangeable protons (see
below). The line width of the exchangeable resonance increases
from ∼70 Hz (20° to 40°C) to ∼150 Hz (54°C) (Figure 1C) which
suggests that the lifetime of the thymidine H-3 imino proton in
the Watson-Crick hydrogen bond decreases with increasing temperature on approaching the melting transition region (midpoint =
59.1°C) of poly(dA-dT) in 0.1 M phosphate solution.

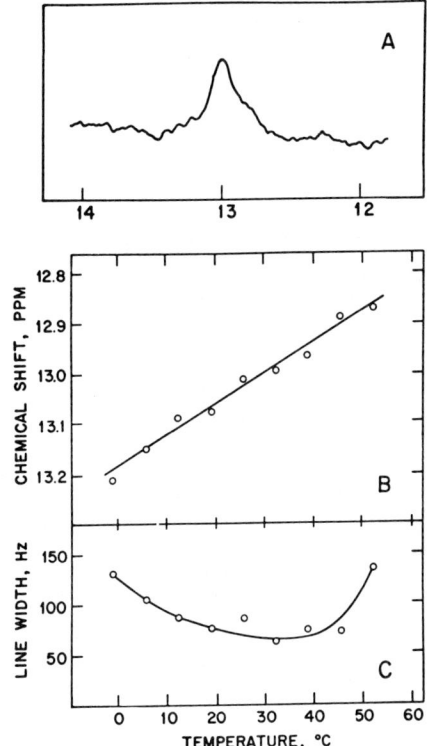

Figure 1. (A) The 360-MHz proton NMR spectrum of the thymidine H-3 proton in poly(dA–dT) in 0.1M phosphate, 1mM EDTA, H_2O at 25.5°C. The (B) chemical shifts and (C) linewidths of this proton in the synthetic DNA in 0.1M phosphate are plotted as a function of temperature between 0° and 55°C. The poly(dA–dT) duplex exhibits a duplex-to-strand transition midpoint of 59°C.

Duplex to Strand Transition: The temperature dependence of
the 360 MHz proton NMR spectra (9 to 4.5 ppm) of poly(dA-dT) in
0.1 M phosphate buffer, ^2H_2O, are presented in Figure 2. Partially
resolvable resonances are observed in the duplex state (48°C) with
line widths of ∿50 Hz and well resolved resonances are observed
in the strand state (67°C) with line widths of a few Hz. The base
resonances can be readily assigned from their chemical shift positions at high temperature while we cannot definitively differentiate between sugar H-1' (and sugar H-3') resonances linked to
the adenosine and the thymidine residues (Figure 2). The duplex
to strand conversion occurs by a cooperative transition at 59°C
with the observable resonances shifting as average peaks during
the melting transition.

The temperature dependence of the chemical shifts of the base
and sugar resonances of poly(dA-dT) in 0.1 M phosphate buffer is
plotted in Figure 3. There are upfield and downfield shifts
associated with the noncooperative premelting transition between
5° and 55°C while only downfield shifts are observed for most of
the base and sugar protons on raising the temperature above 65°C
in the noncooperative postmelting transition temperature range.
The cooperative melting transition (midpoint, $t_{\frac{1}{2}}$ = 59.0°C) exhibits downfield shifts at the base and sugar H-1' protons with
increasing temperature but not at all the remaining sugar protons
(Figure 3).

Resonance Assignments: The adenosine H-2 and thymidine H-6
resonances of poly(dA-dT) exhibit similar chemical shifts around
40°C in 0.1 M phosphate solution (Figure 2). We have evaluated
the spin lattice proton relaxation times, T_1, of the base and
sugar H-1' protons of poly(dA-dT) through the premelting transition and observe that the adenosine H-2 resonance exhibit a T_1
value which is 2 to 3 times longer than that of the remaining
resonances. Thus the adenosine H-2 resonance can be readily
differentiated from the thymidine H-6 resonance since it exhibits
a narrower line width indicative of a longer spin-spin relaxation
time, T_2, and a longer spin-lattice relaxation time T_1 in the
duplex state in solution. The experimental data require that the
adenosine H-2 and the thymidine H-6 resonances cross over at 50°C
(Figure 3).

Further resolution of the 4.5 to 9.0 ppm region of poly(dA-dT) in the duplex state can be achieved by investigating the
related polynucleotide for which bromine replaces the methyl
group at position 5 of the pyrimidine ring. This is readily observable on comparing the spectra of poly(dA-dT) and poly(dA-^{5}brdU) duplexes in the aromatic region at 43°C (Figure 4). The
pyrimidine H-6 proton shifts 0.57 ppm downfield from 7.05 ppm in
poly(dA-dT) to 7.65 ppm in poly(dA-^{5}brdU) at this temperature
(Figure 4). A similar downfield shift of 0.68 ppm is observed at
the pyrimidine H-3 exchangeable proton which shifts from 12.95 ppm
in poly(dA-dT) to 13.65 ppm in poly(dA-^{5}brdU) at 37°C.

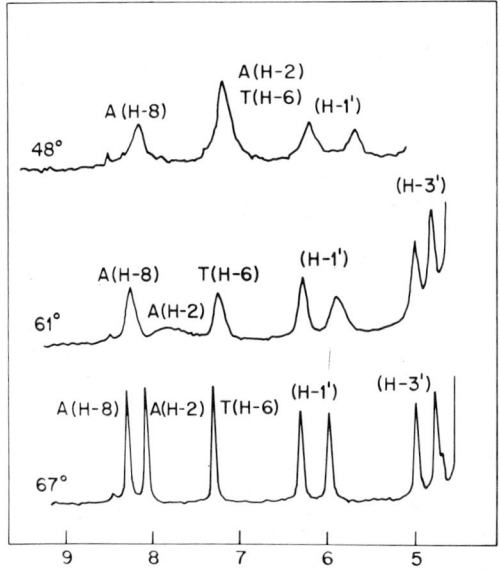

Figure 2. The 360-MHz proton NMR spectra (4.5 to 9.0 ppm) of poly(dA–dT) in 0.1M phosphate, 1mM EDTA, ^2H_2O, pH 6.3 in the duplex state (48°C), during the melting transition (61°C) and in the strand state (67°C)

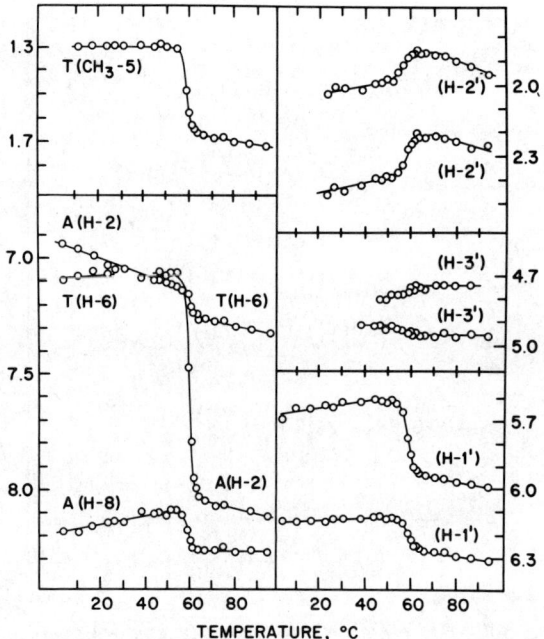

Figure 3. The temperature dependence (5° to 95°C) of the base and sugar proton chemical shifts of poly(dA–dT) in 0.1M phosphate, 1mM EDTA, ^2H_2O, pH 7.0

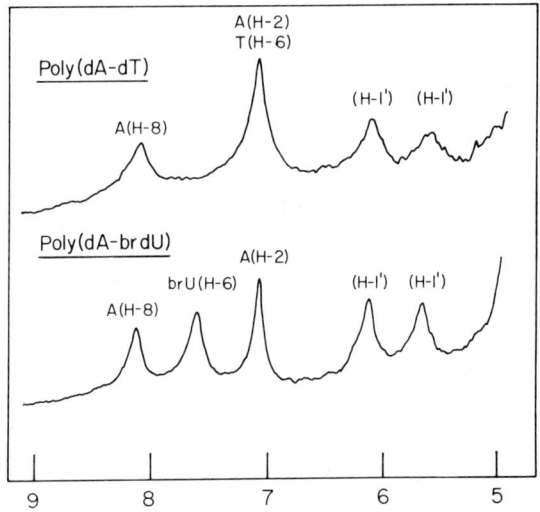

Figure 4. The 360-MHz proton NMR spectra of poly(dA–dT) in 0.1M phosphate, 1mM EDTA, ^2H_2O, pH 7, 43°C and poly(dA–^{5}brdU) in 0.1M NaCl, 10mM phosphate, ^2H_2O, pH 8.1, 43°C

The temperature dependence of the chemical shifts of the base resonances in poly(dA-dT) and poly(dA-^{5}brdU) are plotted in Figure 5. These data demonstrate that the adenosine H-8 and H-2 protons exhibit very similar behavior over the entire temperature range and are not perturbed by the substitution on the pyrimidine 5 position.

Base Pair Overlaps: The chemical shifts in the duplex state, δ_d, and the chemical shift difference between duplex and strand states, $\Delta\delta$, at $t_{1/2}$ of the melting transition, (following correction of the temperature dependent chemical shifts associated with the premelting and postmelting transitions) are summarized in Table I.

The adenosine and thymidine base protons shift upfield to different extents on poly(dA-dT) duplex formation (Table I, Figure 3). These upfield shifts reflect the base pair overlap geometries in the duplex state and result predominantly from ring current contributions due to nearest and next-nearest neighbor base pairs (33,34). These contributions can be computed for the B-DNA overlap geometry and are compared with the experimental upfield chemical shifts on duplex formation of poly(dA-dT) and the related synthetic DNA poly(dA-dU) in Table II.

There is good agreement between the experimental and calculated values at the adenosine H-2, uridine H-5 and H-3 positions for which the experimental chemical shift changes are >0.5 ppm (Table II). By contrast, the agreement is poor at the adenosine H-8 and pyrimidine H-6 positions, for which small experimental shifts were observed (Table II). The adenosine H-8 and pyrimidine H-6 protons are directed towards the sugar-phosphate backbone and factors in addition to the ring current effects may contribute to the observed shifts.

Duplex Dissociation Rates: The adenosine H-8 resonances exhibit a line width of ~50 Hz at 55°C, shift as an average resonance during the melting transition of poly(dA-dT) in 0.1 M phosphate ($\Delta\delta$ at $t_{1/2}$ = 0.184 ppm) and narrow to their value in the strand state above 65°C (Figure 6). By contrast, the adenosine H-2 resonance ($\Delta\delta$ at $t_{1/2}$ = 0.899 ppm) exhibits uncertainty broadening contributions in the fast exchange region during the melting transition, with observed line widths of >100 Hz at the midpoint of the transition (Figure 2). The excess line width contribution to the adenosine H-2 resonance (relative to the adenosine H-8 resonance) at a given temperature during the melting transition, can be combined with the chemical shift difference associated with the melting transition, $\Delta\nu$, in Hz, and the population of duplex (f_d) and strand ($f_s = 1-f_d$) states, to yield the dissociation rate constant, k_d, (= τ_d^{-1}), for conversion from duplex to strands

$$\text{excess width} = 4\pi f_s^2 f_d^2 (\Delta\nu)^2 (\tau_s + \tau_d).$$

The calculated dissociation rate constants for poly(dA-dT) in 0.1 M phosphate buffer solution at f_d = 0.47, 0.24 and 0.009 are presented in Table III with magnitudes of ~1.5 × 10^3 sec^{-1} in the vicinity of the midpoint of the melting transition.

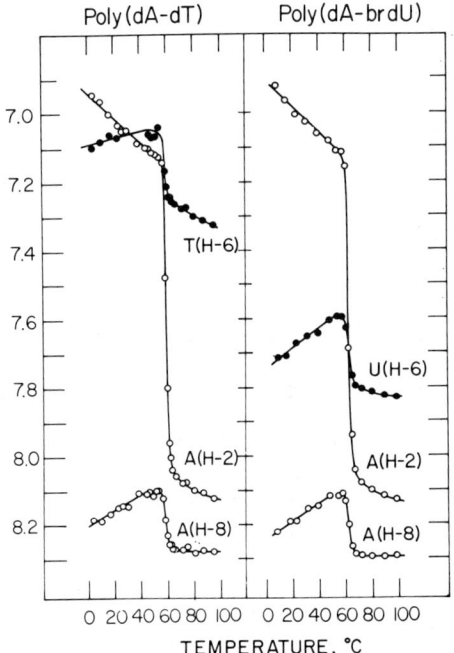

Figure 5. The temperature dependence of the adenosine H-2, pyrimidine H-6, and adenosine H-8 protons in poly(dA–dT) in 0.1M phosphate, 1mM EDTA, ^2H_2O, pH 7, and poly(dA–^{5}brdU) in 0.1M NaCl, 10mM phosphate, 1mM EDTA, ^2H_2O, pH 8.1

TABLE I

Chemical Shift Parameters Associated With The Melting Transition ($t_{\frac{1}{2}}$ = 59°C) of Poly(dA-dT) in 0.1 M Phosphate Solution

	δ_d, ppm [a]	$\Delta\delta$, ppm [b]
A(H-8)	8.085	0.184
A(H-2)	7.139	0.899
T(H-6)	7.052	0.194
T(CH$_3$-5)	1.300	0.359
u(H-1')[c]	5.592	0.343
d(H-1')[c]	6.118	0.148

[a] Chemical shift in the duplex state, δ_d, is defined as the extrapolation of the temperature dependent premelting shift to its value at the transition midpoint.

[b] The duplex to strand transition chemical shift change $\Delta\delta$, is defined as the chemical shift difference following extrapolation of the temperature dependent premelting and postmelting shifts to their values at the transition midpoints. The $\Delta\delta$ values only approximate the total shift change on proceeding from a stacked duplex to unstacked strands.

[c] u(H-1') and d(H-1') represent the upfield and downfield sugar protons respectively.

TABLE II

A Comparison of the Experimental And
Calculated Upfield Shifts Associated With
Poly(dA-dT) and Poly(dA-dU) Duplex Formation

	Experimental[a]		Calculated[b]
	poly(dA-dU)	poly(dA-dT)	poly(dA-dU)
A(H-8)	0.17	0.18	0.04
A(H-2)	0.92	0.90	1.06
U/T(H-6)	0.34	0.19	0.06
U/T(H-3)	1.9[c]	1.7[c]	1.88
U(H-5)	0.68		0.62

[a] The strand to duplex transition upfield chemical shift change $\Delta\delta$, is defined as the chemical shift difference following extrapolation of the temperature dependent premelting and postmelting shifts to their values at the transition midpoints. These $\Delta\delta$ values only approximate the total shift change on proceeding from a stacked duplex to unstacked strands.

[b] The computed upfield ring current shifts are based on nearest neighbor, next-nearest neighbor and cross-strand ring current contributions tabulated by Arter and Schmidt.(34)

[c] The experimental pyrimidine H-3 upfield shift represents the difference between the observed resonance chemical shift in the polynucleotide duplex extrapolated to the transition midpoint and the 14.7 ppm intrinsic position for an isolated base pair.

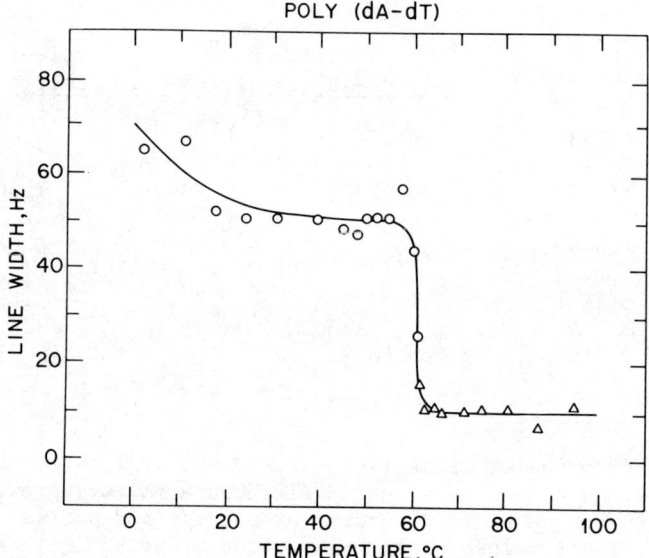

Figure 6. The temperature dependence (5° to 95°C) of the downfield H-1' (△) and the adenosine H-8 (○) linewidths of poly(dA–dT) in 0.1M phosphate, 1mM EDTA, ^2H_2O, pH 7.0

By contrast, the synthetic 30 base pair DNA block polymer duplex $(dC_{15}dA_{15}) \cdot (dT_{15}dG_{15})$ exhibits slow exchange between duplex and strand states during the melting transition (35). The dissociation rate for DNA denaturation is predicted to be proportional to the inverse square of the length of the duplex to be unwound and for molecular weights of 10^6 would be several orders of magnitude slower (36) than the $\sim 10^3$ sec^{-1} rate constants evaluated for the poly(dA-dT) melting transition in solution. The NMR results support the earlier conclusions of Baldwin and coworkers that poly(dA-dT) unfolds by the opening of short branched duplex regions which melt independently of each other (16,17).

Premelting Transition (37,38): The nucleic acid base and sugar resonances of poly(dA-dT) (Figures 3 and 5) and poly(dA-^{5}brdU) (Figure 5) exhibit non-cooperative chemical shift changes with temperature in the duplex state. Both upfield and downfield shifts are observed during this premelting transition at the non-exchangeable protons (Figures 3 and 5) while an upfield shift is observed at the exchangeable thymidine H-3 proton (Figure 1) with increasing temperature.

The observed premelting transition may reflect a conformational change in the polynucleotide duplex with temperature. The adenosine H-2 resonance located in the minor groove shifts to high field while the thymidine CH$_3$-5 group located in the major groove remains unchanged on lowering the temperature in the premelting transition region (Figure 3). This suggests that the base pairs may tilt relative to each other on raising the temperature in the premelting region with a kink site opening into the minor groove (39). The adenosine H-8 and uridine H-6 protons which are directed towards the phosphodiester backbone, and the two sugar H-1' protons, which are sensitive to variations in the glycosidic torsion angles, all shift downfield with decreasing temperature (Figure 3). This suggest that the premelting transition may also involve conformational changes in the sugar-phosphate backbone and glycosidic torsion angles.

Baldwin and coworkers have demonstrated that the degree of branching in the poly(dA-dT) duplex varies significantly with temperature (16,17). Therefore, the premelting transition chemical shift changes may also reflect, in part, the conversion from a highly branched duplex at temperatures just below the melting transition (40) to a less branched duplex structure on lowering the temperature.

There is no information to our knowledge on premelting transitions in RNA duplexes. We have therefore investigated the NMR parameters for the related synthetic RNA duplex poly(A-U) as a function of temperature. Typical 360 MHz aromatic proton region spectra for the poly(A-U) duplex in 0.1 M phosphate ($t_{\frac{1}{2}}$ = 66.5°C) at 8.2° and 44.5°C are presented in Figure 7. It is clear that the chemical shifts of the base protons vary with temperature in the duplex state (Figure 7) and the results are plotted in Figure 8. It may be noted that the direction of the premelting changes at individual resonances exhibit the same behavior in the synthetic DNA (Figure 3) and the synthetic RNA (Figure 7). This suggests that the origin of the premelting conformation (37,38) is common for synthetic RNA and DNA polynucleotide duplexes with the same alternating purine-pyrimidine sequence.

Phosphodiester Linkage: The sugar-phosphate backbone of nucleic acids can be probed at the phosphodiester linkage by ^{31}P NMR spectroscopy. There are only two kinds of phosphodiester linkages in alternating purine-pyrimidine polynucleotides, namely dApdT and dTpdA in poly(dA-dT).

The 145.7 MHz proton noise decoupled ^{31}P NMR spectra of the poly(dA-dT) duplex in 10 mM cacodylate buffer between 28° and 54°C are presented in Figure 9. A broad symmetrical unresolved resonance is observed at 28°C. By contrast, two resolved narrow resonances separated by ~0.2 ppm have been observed for 150 base pair long (dA-dT)$_n$ (41). Thus, though the resolution of dTpdA and dApdT phosphodiesters cannot be achieved at the synthetic DNA level in solution (18), it has been observed for the same sequence at a shorter well defined length (41). More recently, two resolved ^{31}P resonances have also been reported in poly(dA-dT) fibers oriented parallel to the direction to the magnetic field by solid state ^{31}P NMR spectroscopy (42).

The ^{31}P spectra of poly(dA-dT) exhibit some interesting spectral changes in the premelting region in solution. The resonance(s) of poly(dA-dT) exhibit an assymmetric shape with increasing temperature with the component to higher field remaining broad compared to the component to lower field, which narrows considerably at 54°C (Figure 9). This suggests that there exist regions in the polynucleotide duplex with increased segmental mobility in the premelting transition region.

The phosphodiester resonance moves downfield and sharpens dramatically during the melting transition and continues to shift downfield with increasing temperature in the postmelting transition region (18).

Summary: The above results convincingly demonstrate that high resolution proton NMR spectra can be recorded for high molecular weight poly(dA-dT) in aqueous solution. The base pairing in the duplex state has been verified by monitoring the ring imino exchangeable protons in H$_2$O solution. The overlap of adjacent base pairs is readily demonstrated by the observation of upfield shifts on duplex formation and some estimate of the overlap geometries evaluated from the relative magnitudes of the changes at the different proton markers. The observation of average resonances for the nonexchangeable protons during the duplex to strand transition requires fast interconversion between states on the NMR time scale consistent with earlier suggestions that poly(dA-dT) melts by opening of shorter branched duplex regions.

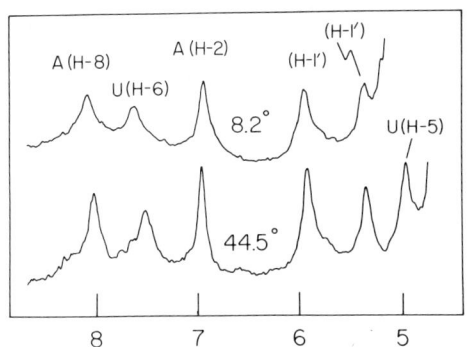

Figure 7. The temperature dependence (8.2°C and 44.5°C) of the 360-MHz proton NMR spectra (4.5 to 9.0 ppm) of poly(A–U) in 0.1M phosphate, 1mM EDTA, ^2H_2O, pH 7.0

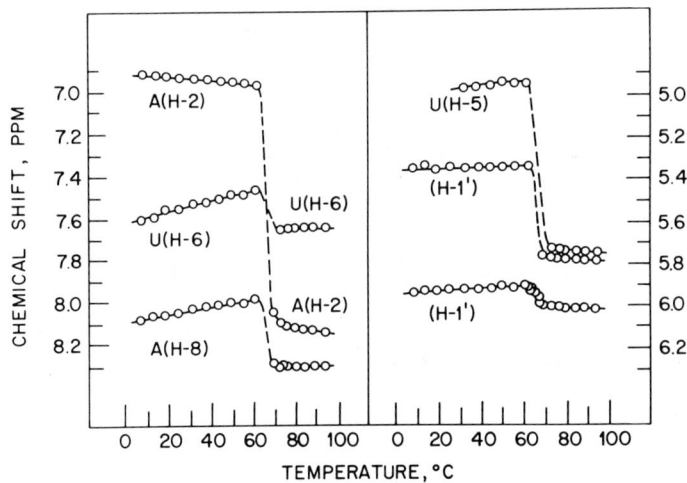

Figure 8. The temperature dependence (5° to 95°C) of the base and sugar H-1' proton chemical shifts of poly(A–U) in 0.1M phosphate, 1mM EDTA, ^2H_2O, pH 7.0

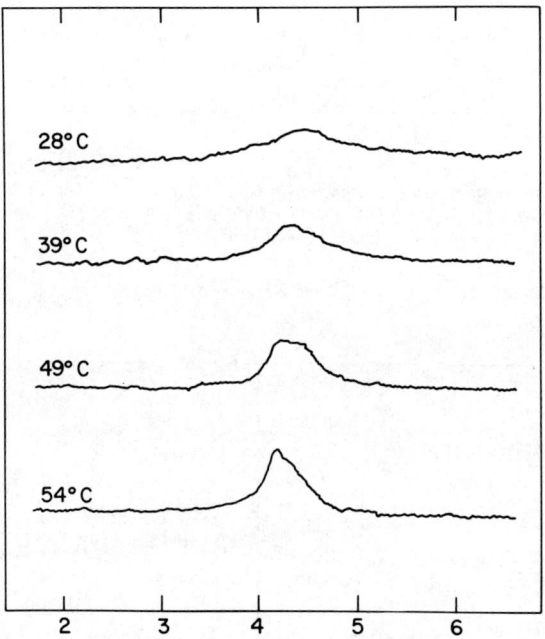

Figure 9. The proton noise decoupled 145.7-MHz ^{31}P NMR spectra of poly(dA–dT) in 0.1M cacodylate, 10mM EDTA, ^2H_2O, pH 7.08 between 28° and 54°C ($t_{1/2}$ of complex is 60°C). The scale is upfield from standard trimethylphosphate.

The premelting conformational transition can be monitored at the chemical shifts of the exchangeable and nonexchangeable protons and the line shape of the phosphodiester ^{31}P resonance of poly(dA-dT) and its related synthetic DNA's in solution. The results suggest that the base pairs may partially unstack into the minor groove, accompanied by small changes in the glycosidic torsion angles and the sugar-phosphate backbone as the temperature is raised in the premelting transition region.

Counterion Binding. Alkylammonium Ions

The research below focusses on the NMR parameters for poly(dA-dT) in 1 M tetramethylammonium chloride (TMA$^+$) relative to their value in the same concentration of sodium chloride. The methyl groups shield the charged nitrogen in the TMA$^+$ ion and it was of interest to determine whether conformational changes occur in the synthetic DNA when the counterion was changed from Na$^+$ to TMA$^+$.

The NMR experiments were undertaken on 28 mM (in nucleotides) poly(dA-dT) in 10 mM cacodylate buffer, to which 1 M salt solutions were added. The samples contained in addition 1 mM and 10 mM EDTA for the proton and phosphorus NMR studies, respectively. The experimental conditions are indicated in order to emphasize that even in 1 M TMA$^+$ solutions there are Na$^+$ ions associated with poly(dA-dT), the 10 mM buffer and the EDTA solutions.

Equilibrium dialysis studies on the relative affinities of weakly bound cations with varying base composition DNA have demonstrated that tetraalkylammonium ions bind more tightly to dA·dT rich DNA compared to dG·dC rich DNA (43).

Hydrogen Bonding: The thymidine H-3 proton of poly(dA-dT) can be readily monitored in 1 M TMA$^+$ solutions at pH 7.5 and pH 9.5 at 37°C (Figure 10). This demonstrates that the base pairs are intact in the synthetic DNA and that their exchange is not base catalyzed up to pH 9.5 at 37°C. These results parallel observations on poly(dA-dT) with Na$^+$ as the counterion and in poly(dA-dT) complexes with intercalating agents. By contrast, in steroid diamine·poly(dA-dT) complexes which involve tilted base pairs at the binding site, the exchange of the thymidine H-3 protons are susceptible to base catalysis (44).

The line width of the thymidine H-3 proton is compared with the corresponding value for the adenosine H-8 nonexchangeable proton for poly(dA-dT) in 1 M TMA$^+$ solution, pH 7.5, in Figure 11. The exchangeable resonance exhibits the larger line width in the duplex state due to the interaction of its proton with the directly bonded ^{14}N quadrupolar nucleus. The exchangeable resonance broadens dramatically between 67.5°C (line width ∿150 Hz) and 77.5°C (line width ∿250 Hz), a temperature region where the duplex is still intact from the nonexchangeable proton data. These

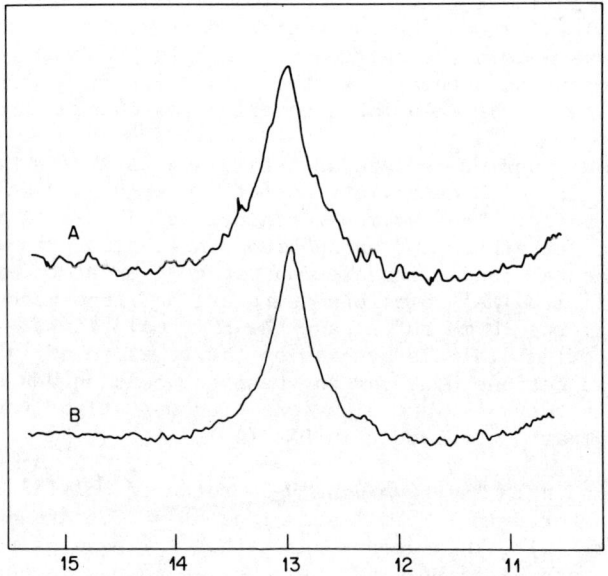

Figure 10. The 360-MHz correlation proton NMR spectra of poly(dA–dT) in 1M ($^2H_3C)_4NCl$, 10mM phosphate, 1mM EDTA, 80% H_2O–20% D_2O at 37°C. Spectrum A was recorded at pH 7.5 while Spectrum B was recorded at pH 9.5. ($^2H_3C)_4NCl$ was purchased from Merck and used without further purification.

results suggest that exchange occurs by transient opening of the duplex at temperatures just below the onset of the melting transition.

The temperature dependences of the thymidine H-3 protons for poly(dA-dT) in 1 M Na^+ and 1 M TMA^+ are compared in Figure 12. The chemical shifts are similar at high temperature and differ by ~0.1 ppm at the lower temperature (Figure 12). This suggests that similar base pair overlaps are observed for poly(dA-dT) in 1 M Na^+ and 1 M TMA^+ as monitored at the thymidine H-3 proton located in the center of the base pair.

Glycosidic Torsion Angle: Additional structural information can be deduced from a comparison of the nonexchangeable proton chemical shifts during the helix-coil transition of poly(dA-dT) in 1 M Na^+ and 1 M TMA^+ solutions. The temperature dependent chemical shifts in the premelting and melting transitions are plotted in Figure 13. The adenosine (H-8 and H-2) and thymidine (H-6 and CH_3-5) base protons exhibit similar duplex chemical shifts in Na^+ and TMA^+ solutions, indicative of similar base pair overlap geometries in the double helix. By contrast, a comparison of the temperature dependent H-1' and H-3' protons demonstrates a selective upfield shift at the H-1' resonance at higher field on proceeding from Na^+ to TMA^+ solution (Figure 13). Further, the direction of the premelting transition chemical shift change is different at this H-1' resonance under the two salt conditions (Figure 13). The sugar H-1' proton chemical shift is sensitive to changes in the glycosidic torsion angle and the NMR results suggest that the TMA^+ ion induces a torsion angle change at either the adenosine or thymidine glycosidic bond.

Melting Transition: The poly(dA-dT) duplex is stabilized by the tetraalkylammonium ion relative to the sodium ion since the melting transition midpoints of the synthetic DNA are 79°C in 1 M TMA^+ solution and 72.5°C in 1 M Na^+ solution (Figure 13).

The adenosine H-2 resonance of poly(dA-dT) broadens during the melting transition in 1 M TMA^+ solution (Figure 14) and the excess width due to uncertainty broadening gives a dissociation rate constant $k_d = 2.0 \times 10^3$ sec^{-1} at 78.8°C ($t_{\frac{1}{2}} = 79.0°C$). A similar value is observed for the rate constant when 1 M Na^+ is the cation ($k_d = 2.3 \times 10^3$ sec^{-1} at 71.9°C; $t_{\frac{1}{2}} = 72.5°C$) so that the duplex dissociation rates of the synthetic DNA are similar in Na^+ and TMA^+ solution.

Phosphodiester Linkages: The dTpdA and dApdT phosphodiester linkages do not give resolved ^{31}P spectra for poly(dA-dT) in 1 M NaCl solution (10). It is therefore striking that two partially resolved resonances are observed for this synthetic DNA in 1 M TMA^+ solution (Figure 15). The chemical shift separation was 0.42 ppm at 52°C and 0.34 ppm at 67°C. Partially resolved ^{31}P resonances have been reported for 150 base pair $(dA-dT)_n$ in the

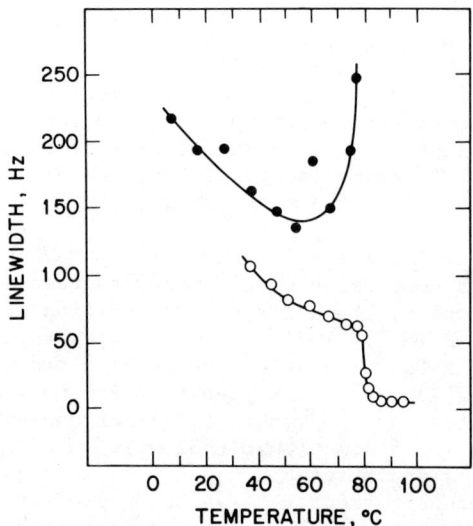

Figure 11. A comparison of the temperature dependence of the linewidth of the nonexchangeable adenosine H-8 resonance (○) and the exchangeable thymidine H-3 resonance (●) in 1M ($^2H_3C)_4NCl$, 10mM phosphate, 1mM EDTA, aqueous solution, pH ~ 7.5

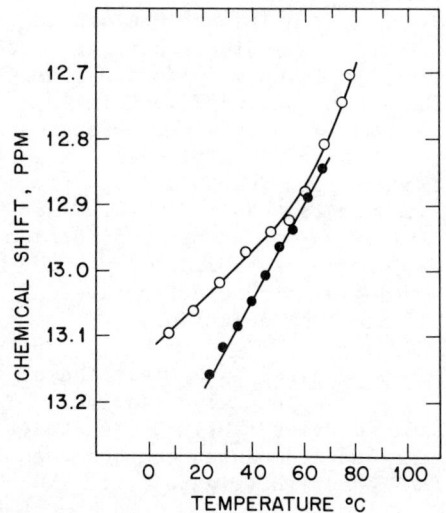

Figure 12. A comparison of the temperature dependence of the thymidine H-3 proton chemical shift of poly(dA-dT) in 1M NaCl, 10mM cacodylate, 0.1mM EDTA, H_2O, pH 6.53 (●) and in 1M ($^2H_3C)_4NCl$, 10mM phosphate, 1mM EDTA, H_2O, pH 7.5 (○)

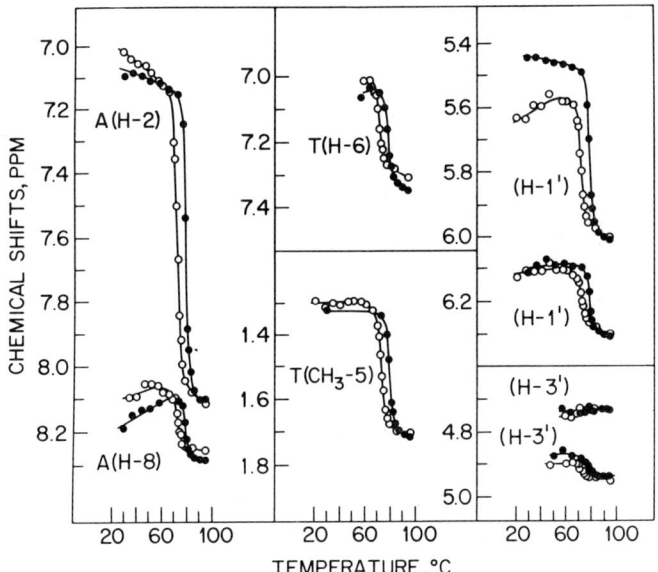

Figure 13. The temperature dependence of the base and sugar proton resonances of poly(dA–dT) in 1M NaCl, 10mM cacodylate solution (○), and in 1M (^2H_3C)$_4$-NCl, 10mM phosphate solution (●). (^2H_3C)$_4$NCl was purchased from Merck and used without further purification.

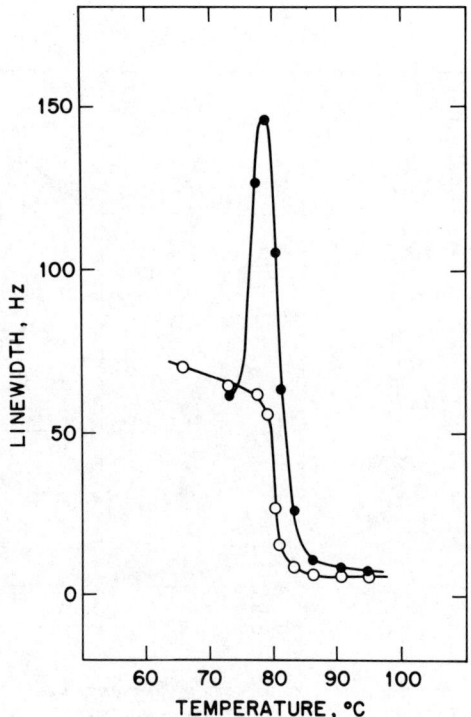

Figure 14. A comparison of the temperature dependence of the linewidth of the adenosine H-8 resonance (○) and the adenosine H-2 resonance (●) in 1M $(^2H_3C)_4NCl$, 10mM phosphate, 1mM EDTA, ^2H_2O

absence of TMA⁺ but the reported shift difference of ∿0.25 ppm (41) is smaller than the results reported in this review in TMA⁺ solution. The two ^{31}P resonances should exhibit equal area if one corresponds to the dTpdA and the other to the dApdT phosphodiester linkages. We do not understand the origin of the unequal areas in the ^{31}P spectrum of poly(dA-dT) in TMA⁺ solution with the relative ratios maintained in 2 M and 4 M salt.

The ^{31}P NMR results suggests that there may be small differences between the purine(3'-5')pyrimidine and pyrimidine-(3'-5')purine phosphodiester linkages in synthetic DNA's with an alternating sequence in 1 M TMA⁺ solution.

Summary: The thymidine H-3 hydrogen-bonded proton is well shielded from solvent in the premelting transition region of poly(dA-dT) in 1 M TMA⁺ solution based on an analysis of its line width as a function of pH and temperature. This result is consistent with parallel stacking of base pairs in the duplex state of the synthetic DNA in TMA⁺ solution.

We observe similar base pair overlaps for poly(dA-dT) in 1 M Na⁺ and TMA⁺ solution based on a comparison of the base proton chemical shifts of the synthetic DNA in the duplex state as a function of counterion.

The tetramethylammonium ion appears to exhibit some specificity in its binding to poly(dA-dT) since the dTpdA and dApdT resonances are partially resolved in the presence of this counterion. This specificity is probably also reflected in the selective glycosidic torsion angle change (monitored at one sugar H-1' resonance) in the poly(dA-dT) duplex on proceeding from 1 M Na⁺ solution to 1 M TMA⁺ solution.

Proflavine-DNA Complexes

Intercalation of planar molecules between base pairs was first postulated by Lerman based on spectroscopic and hydrodynamic investigations of the binding of acridine dyes with duplex DNA (45,46). This intercalation model received early support when Waring demonstrated that the mutagen proflavine unwound covalently circular superhelical DNA (47). Definitive evidence for intercalation resulted from single crystal X-ray studies of the 3:2 proflavine:C-G complex (48), 2:2 proflavine:iodoC-G complex (49), and the 2:2 proflavine:dC-dG complex (50) where a proflavine molecule is intercalated between the base pairs of a Watson-Crick miniature double helix. Arnott and corworkers have utilized linked atom conformational calculations to define the conformational characteristics at the binding site of proflavine intercalated into DNA (51) and RNA (52). The NMR investigation on proflavine·synthetic DNA complexes reported below (25) complement the results obtained from the crystallographic studies on the complex at the dinucleotide level (48-50) and modelling investigations at the polynucleotide level. (51,52) [3]

Planar dyes such as proflavine can bind to DNA by intercalation between base pairs or by stacking along the groove of the helix. The former process is favored when the nucleic acid is in excess (Nuc/D $\gtrsim$ 4) and in high salt solution. We are interested in the intercalation process and hence our NMR studies have focussed on an investigation of the proflavine·poly(dA-dT) complex as a function of the Nuc/D ratio in 1 M NaCl in an attempt to probe the structure and dynamics of mutagen-nucleic acid interactions in solution (25).

Hydrogen Bonding: The thymidine H-3 Watson-Crick proton can be readily detected in the proflavine·poly(dA-dT) complex, Nuc/D = 8, in 1 M NaCl solution with the resonance shifting to high field on complex formation (Figure 16). The results demonstrate that the base pairs are intact in the proflavine complex with the synthetic DNA.

Melting Transition: Typical 360 MHz proton NMR spectra of proflavine·poly(dA-dT) complexes, Nuc/D = 24 and Nuc/D = 8, in 1 M NaCl solution at temperatures below the midpoint for the dissociation of the complex are presented in Figures 17A and B respectively. The stronger base and sugar resonances can be readily resolved from the weaker proflavine resonances (designated by asterisks) in the presence of excess nucleic acid (Figure 17) so that the resonances of the synthetic DNA and the mutagen can be monitored independently of each other.

The dissociation of the proflavine·poly(dA-dT) complex can be followed by monitoring the temperature dependent chemical shift or the line width as demonstrated by shift data on the thymidine CH_3-5 resonance (Figure 18A) and width data on the adenosine H-8 resonance (Figure 18B). The proton resonances shift as average peaks during the dissociation of the complex, indicative of fast exchange ($k_{dissociation} \sim 10^4$ sec^{-1} at the transition midpoint) between the complex and its dissociated components on the NMR time scale.

The transition midpoint increases from 72.6°C for poly(dA-dT) in 1 M NaCl to 78.1°C for the Nuc/D = 24 proflavine complex to 83.4°C for the Nuc/D = 8 proflavine complex, indicative of stabilization of the duplex by bound mutagen.

Nucleic Acid Base Resonances: The chemical shifts of the nonexchangeable protons in poly(dA-dT), the Nuc/D = 24 complex and the Nuc/D = 8 complex in 1 M NaCl solution are plotted as a function of temperature in Figure 19. The nucleic acid nonexchangeable proton chemical shifts in the duplex state are either unperturbed (adenosine H-8, H-2, and thymidine CH_3-5) or shift slightly upfield (thymidine H-6) on complex formation (Figure 19). By contrast, the thymidine H-3 exchangeable proton located in the center of the duplex resonates $\sim$0.35 ppm to higher field in the Nuc/D = 8 proflavine complex compared to its position in the

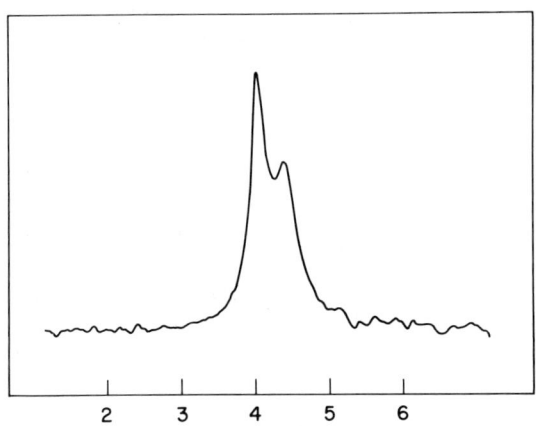

Figure 15. The proton noise decoupled 145.7-MHz ^{31}P NMR spectra of poly-(dA–dT) in 1M $(CH_3)_4NCl$, 10mM cacodylate, 1mM EDTA, ^2H_2O, pH 7.95 at 67°C. The chemical shifts are upfield from standard trimethylphosphate.

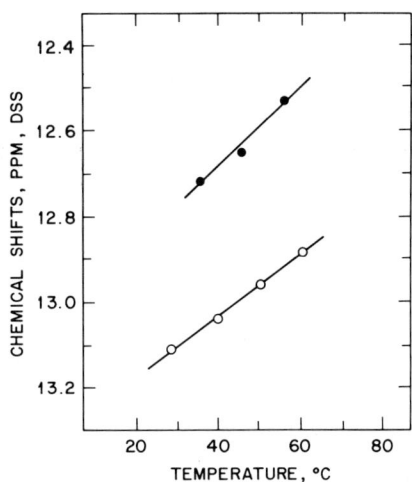

Figure 16. The temperature dependence of the thymidine H-3 resonance in poly(dA–dT) (○) and the proflavine · poly(dA–dT) complex Nuc/D = 8 (●) in 1M NaCl, 10mM cacodylate, 1mM EDTA, H_2O at pH 6.53, and pH 7.1 respectively

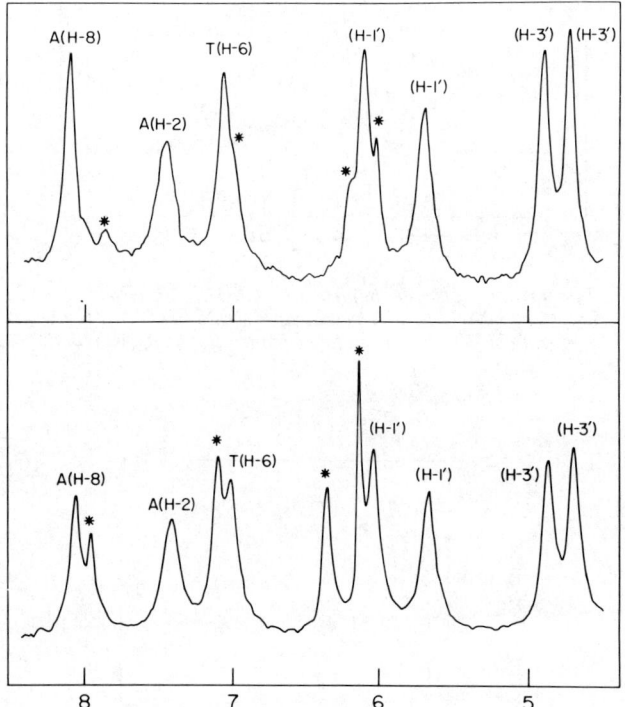

Figure 17. The 360-MHz proton NMR spectra of the proflavine · poly(dA–dT) complex in 1M NaCl, 10mM cacodylate, 10mM EDTA, ^2H_2O, pH 7. The top spectrum represents the Nuc/D = 24 complex at 78.5°C ($t_{1/2}$ of the proflavine resonances in the complex is 80°C), while the bottom spectrum represents the Nuc/D = 8 complex at 81.4°C ($t_{1/2}$ of proflavine resonances in complex is 84.3°C). The proflavine resonances are designated by asterisks.

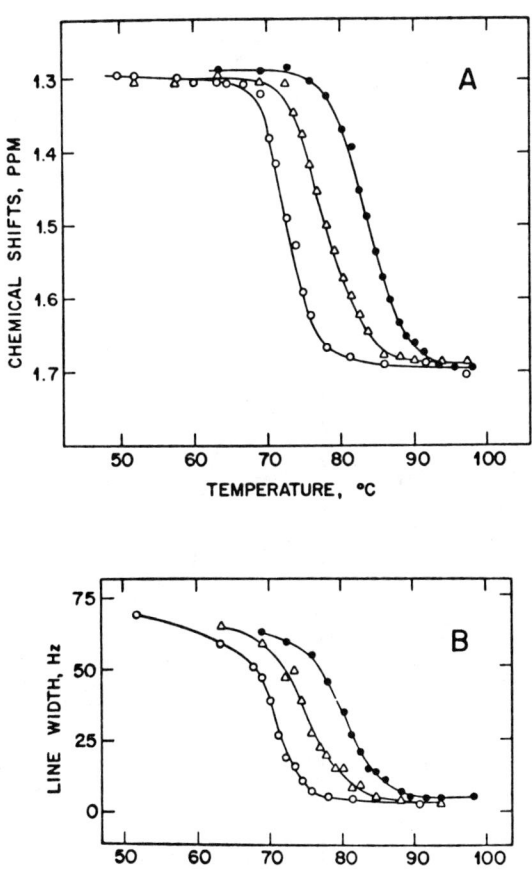

Figure 18. The temperature dependence of (A) the thymidine CH_3-5 chemical shift and (B) the adenosine H-8 linewidth in poly(dA–dT) (○), the proflavine · poly(dA–dT) complex, Nuc/D =24 (△) and Nuc/D = 8 (●) in 1M NaCl, 10mM cacodylate, 10mM EDTA, ^2H_2O, pH 7

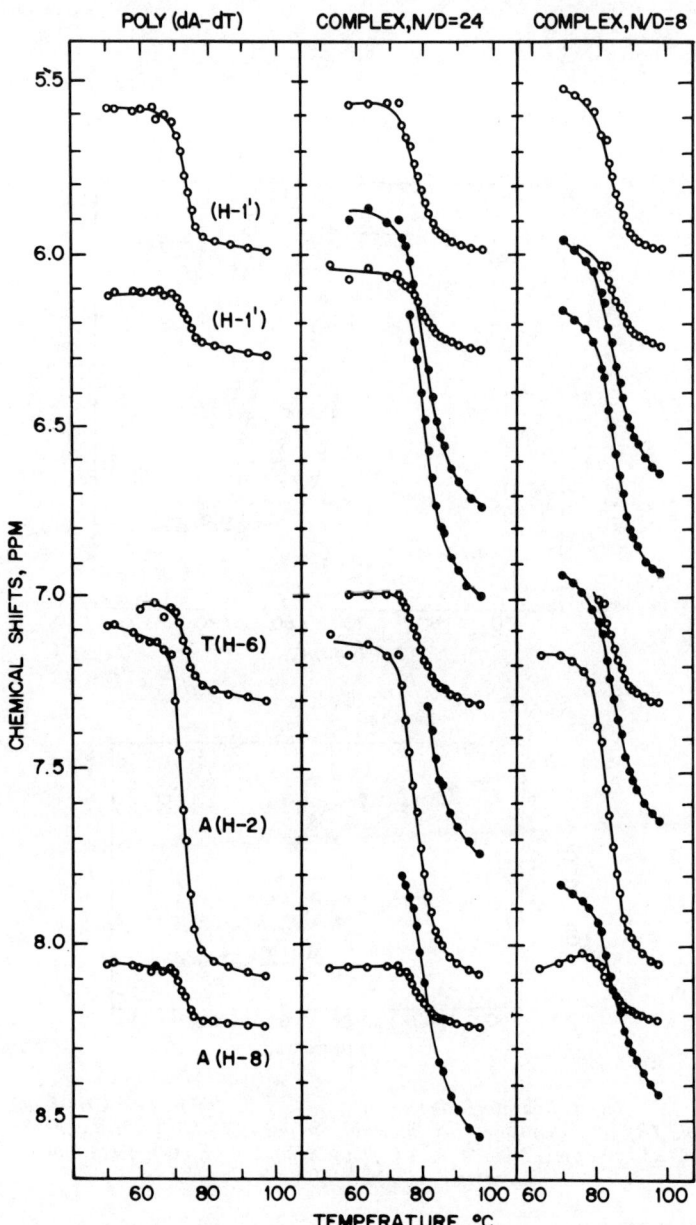

Figure 19. The temperature dependence of the nucleic acid (○) and proflavine (●) chemical shifts between 5.5 and 8.6 ppm for poly(dA–dT) and the Nuc/D = 24 and 8 proflavine · poly(dA–dT) complexes in 1M NaCl, 10mM cacodylate, 10mM EDTA, ^2H_2O between 50° and 100°C. The poly(dA–dT) concentration was fixed at 12.6mM in phosphates and the proflavine concentration was varied to make the different Nuc/D ratio complexes.

synthetic DNA (Figure 16). The nucleic acid complexation shifts reflect the difference in ring current contributions between the proflavine ring (53) and a dA·dT base pair (33,34) which is displaced following intercalation.

Nucleic Acid Sugar Resonances: Decoupling studies have correlated the resolved protons on each of the two sugar rings of poly(dA-dT) (Table IV) though it is not yet possible to definitively differentiate between the adenosine and thymidine residues. Several sugar ring protons can be monitored in the proflavine·poly(dA-dT) complex and these include H-1' position (Figure 19) and H-3' and H-2',2" positions (Figure 20).

Both sugar H-1' protons of poly(dA-dT) undergo small upfield shifts on proflavine complex formation with a somewhat more pronounced effect on the H-1' at lower field, which shifts from 6.1 ppm in the synthetic DNA to 5.95 ppm in the Nuc/D = 8 complex (Figure 19). This latter observation was previously reported for the ethidium bromide·poly(dA-dT) complex (11). The sugar H-1' protons predominantly monitor changes in the glycosidic torsion angle (54) and the results suggest that the generation of the proflavine intercalation site requires changes in these torsion angle(s). The result is consistent with parallel observations on the X-ray structures of intercalative drug-dinucleoside miniature duplexes (48-50).

The sugar H-3' protons of poly(dA-dT) undergo chemical shift changes of <0.1 ppm during the duplex to strand transition (Figures 3 and 20). It is therefore significant that the downfield and upfield H-3' protons shift to high field by 0.05 ppm and 0.1 ppm on formation of the Nuc/D = 8 complex (Figure 20). The sugar H-3' protons are attached to the sugar-phosphate backbone and are sensitive to either the orientation of the sugar ring relative to the helix axis and/or the pucker of the sugar ring. It should be noted that the H-3' shifts on proflavine complex formation (Figure 20) parallel those reported for ethidium bromide complex formation with poly(dA-dT) (11).

Mutagen Resonance: The mutagen resonance in the Nuc/D = 24 and Nuc/D = 8 proflavine·poly(dA-dT) complexes in 1 M NaCl (Figure 19) reflect the free proflavine chemical shifts at high temperature (∿100°C) and proflavine complexed to the synthetic DNA at low temperature (≤60°C). All four proflavine resonances undergo large upfield shifts of ∿0.9 ppm on complex formation (Table V) which can only arise from ring current contributions from adjacent base pairs following intercalation of the mutagen into double helical DNA.

Sequence Specificity: Optical studies on the binding of intercalative drugs to dinucleoside monophosphates demonstrated that actinomycin exhibits a purine(3'-5')pyrimidine specificity (57-60) while ethidium exhibits a pyrimidine(3'-5')purine (61-63)

TABLE III

Poly(dA-dT) Duplex Dissociation Rate Constants in 0.1 M Phosphate Solution

Temp, °C	$f_d{}^a$	Excess Width, Hz[b]	k_d, sec^{-1}
59.4°	0.47	~96	1.5×10^3
60.6°	0.24	~76	2.4×10^3
61.8°	0.009	20.5	4.1×10^3

[a] f_d = fraction of duplex state.

[b] The difference in line width between the adenosine H-2 resonance and the downfield H-1' resonance.

TABLE IV

Sugar Proton Chemical Shifts in the Poly(dA-dT) Duplex in 1 M NaCl Solution[a]

	H-1'	H-2',2"	H-3'
Sugar 1	5.68	1.87, 2.32	4.76
Sugar 2	6.00	2.74, 2.55	4.90

[a] The H-1', H-2',2" and H-3' protons on each sugar ring were correlated by decoupling experiments.

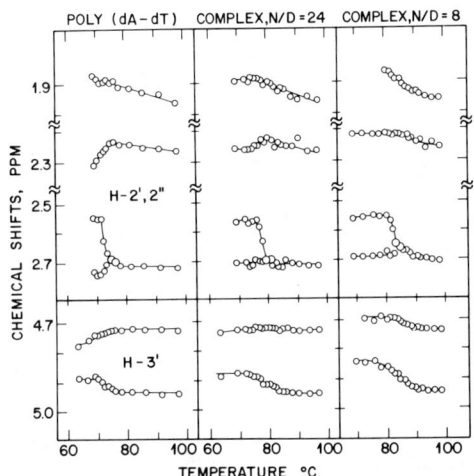

Figure 20. The temperature dependence of the nucleic acid H-3' (4.6 to 5.0 ppm) and the H-2', 2" (1.8 to 2.8 ppm) chemical shifts for poly(dA–dT) and the Nuc/D = 24 and = 8 proflavine · poly(dA–dT) complex in 1M NaCl, 10mM cacodylate, 10mM EDTA, ^2H_2O, pH 7 between 60° and 100°C

TABLE V

Experimental and Calculated Upfield Proflavine
Complexation Shifts on Formation
of the Proflavine·Poly(dA-dT) Complex

Assignment[a]	Experimental Shift, ppm			Calculated Shift, ppm
	Complex[b]	Free[c]	$\Delta\delta$[d]	$\Delta\delta$[e]
H_a	7.820	8.740	0.92	0.85
H_b	6.930	7.880	0.95	0.8
H_c	6.155	7.060	0.905	0.65
H_d	5.955	6.840	0.885	0.7

[a] The proflavine resonance H_a, H_b, H_c, and H_d assignments are as outlined in Reference 25.

[b] Data for the proflavine·poly(dA-dT) complex, Nuc/D = 8 in 1 M NaCl, 10 mM cacodylate, 10 mM EDTA, ^2H_2O, pH 7 at 69°C.

[c] Chemical shift of 1.3 mM proflavine in 0.1 M phosphate, 1 mM EDTA, ^2H_2O, pH 6.6 at 100°C.

[d] The experimental upfield complexation shift is the difference between the values for the intact complex at 69°C relative to free proflavine at 100°C.

[e] The calculations are based on ring current and atomic diamagnetic anisotropy contributions (56) based on the intercalation overlap geometry depicted in the text. The overlap geometry corresponds to intercalation of dT-dA site.

under conditions where the drugs acts as a template on which the nucleic acid forms a miniature intercalative complex. Our laboratory has extended these investigations to the stable oligonucleotide duplex level with NMR studies of the sequence specificity of actinomycin D binding to hexanucleotides (64) and tetranucleotides (65) in aqueous solution.

Tetranucleotides containing dG·dC base pairs form stable duplexes at low temperature so that the self-complementary sequences dC-dC-dG-dG [contains dC(3'-5')dG but no dG(3'-5')dC binding sites] and dG-dG-dC-dC [contains dG(3'-5')dC but no dC(3'-5')dG binding sites] serve as excellent oligonucleotide duplexes for differentiating pyrimidine(3'-5')purine specificity from purine(3'-5')pyrimidine specificity associated with drug complexation (66,67).

We have monitored the decrease in the 435 nm actinomycin D absorbance band on gradual addition of tetranucleotide in 0.1 M phosphate at 24.5°C. A comparison of the tetranucleotide concentrations corresponding to half-maximal change demonstrates stronger binding of actinomycin D to dG-dG-dC-dC compared to dC-dC-dG-dG (Figure 21), establishing a relative sequence specificity for actinomycin D complex formation at purine(3'-5')pyrimidine sites at the oligonucleotide duplex level (67).

By contrast, ethidium bromide (Figure 22) and proflavine (Figure 23) exhibit a relative pyrimidine(3'-5')purine specificity at the duplex level since stronger binding is observed with dC-dC-dG-dG compared to dG-dG-dC-dC in 0.1 M phosphate solution at 1°C.

These results demonstrate that the relative sequence specificity for drug intercalation at the dinucleoside phosphate level (57-63) are also observed at the stable oligonucleotide duplex level (66,67). The latter investigations have been repeated in another laboratory (68,69) with similar conclusions.

Overlap Geometry: A schematic representation of the proposed overlap geometry for proflavine intercalated into a deoxy pyrimidine(3'-5')purine site is presented below with the (o) symbols representing the location of the phenanthridine ring protons. The mutual overlap of the two base pairs at the intercalation site involves features observed in the crystal structures of a platinum metallointercalator·miniature dC-dG duplex complex (55) and the more recent proflavine·miniature dC-dG duplex complex (48), as well as features derived in a linked-atom conformational calculation of the intercalation site in the proflavine·DNA complex (51). [4]

The overlap of proflavine with adjacent base pairs was varied until there was approximate agreement between the experimental upfield complexation shifts (Table V) and those calculated from ring current and atomic diamagnetic anisotropy contributions from the base pairs (56). The calculated upfield shifts are somewhat smaller than the experimental complexation shifts at the proflavine protons in the synthetic DNA complex (Table V). This

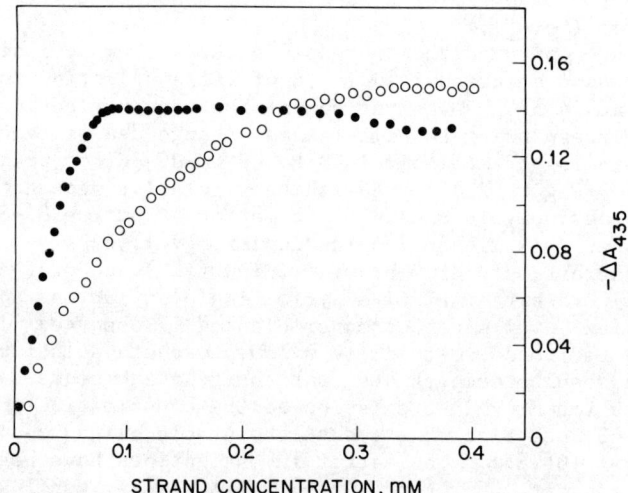

Figure 21. A plot of the change in the 435-nm absorbance of 0.02mM actinomycin D on addition of dC–dC–dG–dG (○) and dG–dG–dC–dC (●) in 0.1M phosphate solution, pH 7.0 at 24.5°C. The tetranucleotide concentrations are based on an extinction coefficient of 2.90×10^4 M^{-1} cm^{-1} for dC–dC–dG–dG (in strands, no added salt, 70°C) and 3.05×10^4 M^{-1} cm^{-1} for dG–dG–dC–dC (in strands, no added salt, 70°C).

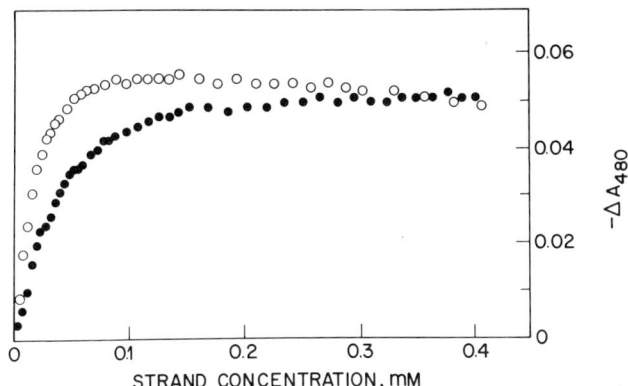

Figure 22. A plot of the change in the 480-nm absorbance of 0.02mM ethidium bromide on addition of dC–dC–dG–dG (○) and dG–dG–dC–dC (●) in 0.1M phosphate solution, pH 7.0 at 1°C

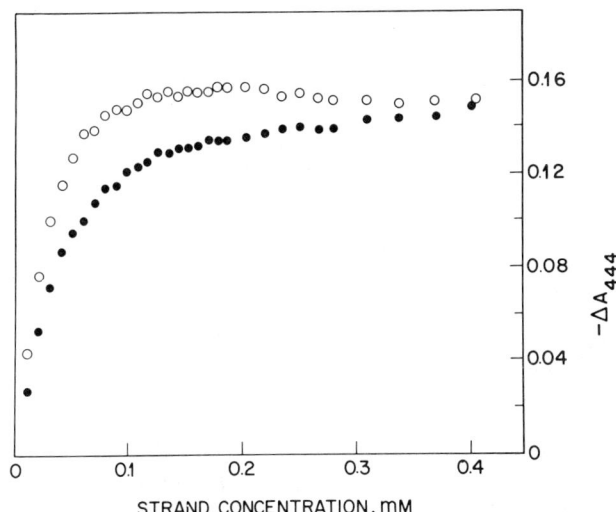

Figure 23. A plot of the change in the 444-nm absorbance of 0.02mM proflavine on addition of dC–dC–dG–dG (○) and dG–dG–dC–dC (●) in 0.1M phosphate solution, pH 7.0 at 1°C

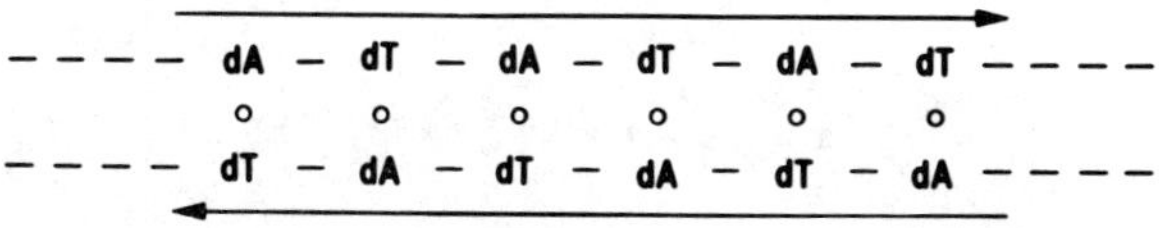
dA
dT
dA
dT
dA
dT
dT
dA
dT
dA
dT
dA

[1]

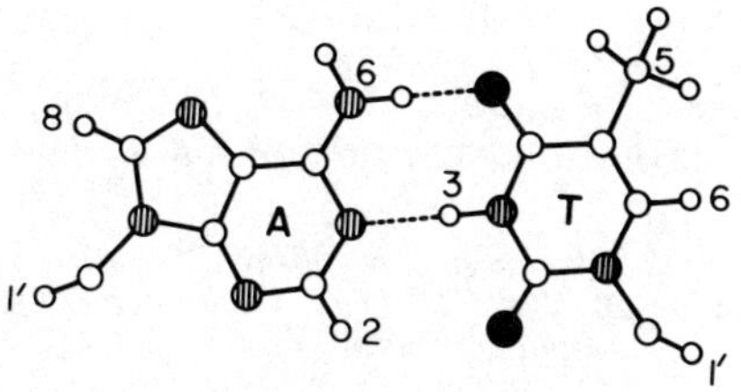
8
6
5
3
A
T
6
1′
2
1′

[2]

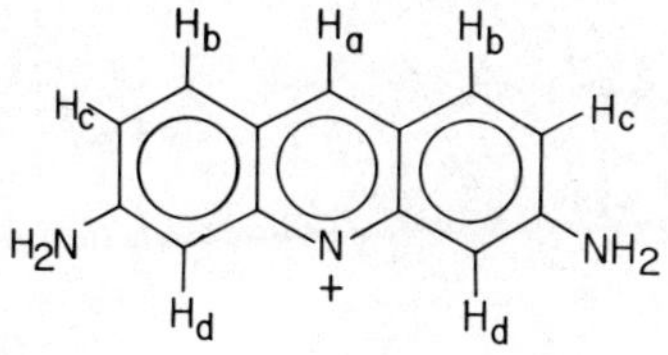
H_b
H_a
H_b
H_c
H_c
H_2N
N
+
NH_2
H_d
H_d

[3]

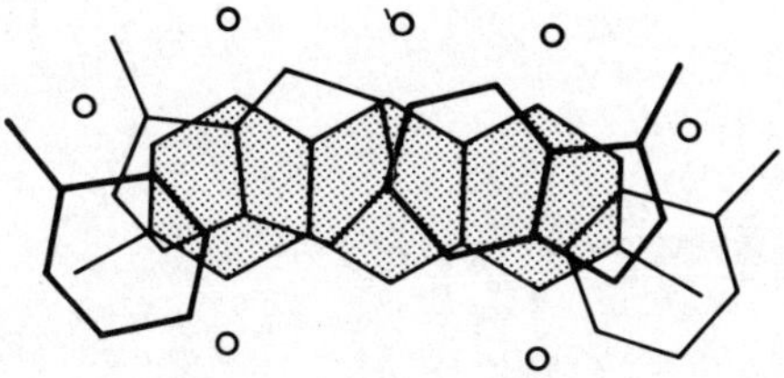

[4]

difference may reflect in part the contributions from next-nearest neighbor base pairs (33) which are not included in the calculation of the proflavine upfield shifts.

The resultant symmetrical overlap geometry at the intercalation site has the long axis of the proflavine ring colinear with the direction of the Watson-Crick hydrogen bonds and there is significant overlap between the mutagen ring system and both purine rings of adjacent base pairs. This overlap geometry deduced from the NMR investigations differs somewhat from that deduced from linked atom conformational calculations (51) in that there is greater overlap between the intercalator and base pairs in the former case.

It should also be stressed that the proflavine is assymmetrically positioned with respect to the base pairs in the proflavine:dC-dG duplex crystal structure (50) but symmetrically placed in the corresponding RNA crystalline complex (48,49).

Phosphodiester Linkage: The dTpdA and dApdT phosphodiester linkages are not resolved in the proton noise decoupled ^{31}P spectrum of the poly(dA-dT) duplex in 1 M NaCl solution (Figure 24A). By contrast, two resolved resonances are observed for the proflavine·poly(dA-dT) complex in the presence of excess nucleic acid with a peak separation of 0.34 ppm in the Nuc/D = 15 complex and 0.42 ppm in the Nuc/D = 10 complex (Figure 24B). One of the resonances in the proflavine complex exhibits a similar chemical shift to that observed in the unresolved envelope of the synthetic DNA while the other resonance shifts downfield as a result of complex formation. The shifted resonance most likely originates from the phosphodiester at the complexation site and may reflect small changes in the phosphodiester O-P torsion angles on generation of the intercalation site (70,71). These downfield shifts in the proflavine·poly(dA-dT) complex are much smaller than the large shifts of 1.5 ppm and 2.5 ppm associated with intercalation of actinomycin into oligonucleotide duplexes (64,65).

Since each resonance is an average of free and mutagen-bound phosphodiester at either the dTpdA or dApdT linkages in the presence of excess nucleic acid, the selective shift of only one of these phosphates suggests that proflavine exhibits a sequence specificity for one of the two binding sites.

Summary: The proton resonances of the nucleic acid and the mutagen are well resolved in the proflavine·poly(dA-dT) complex and can be monitored independently of each other. The resonances shift as average peaks during the thermal dissociation of the complex with stabilization of the duplex by bound mutagen.

The relative magnitude of the large proflavine upfield complexation shifts requires that the dye intercalate into the duplex with its long axis colinear to the direction of the Watson-Crick hydrogen bonds of adjacent base pairs. This results in significant overlap of the proflavine ring system and base pairs at the intercalation site.

We observe structural changes in the glycosidic torsion angle(s) (monitored at the H-1' protons) and the sugar ring parameters (monitored at the H-3' protons) of poly(dA-dT) on proflavine complex formation.

The dye exhibits different binding affinities for pyrimidine(3'-5')purine and purine(3'-5')pyrimidine sites and this is readily demonstrated by the observation of resolved resonances in the ^{31}P spectrum of the proflavine·poly(dA-dT) complex in solution.

Daunomycin-DNA Complexes

Daunomycin and its analog adriamycin are in clinical use as potent antitumor agents in combination chemotherapy against acute lymphocytic leukemia. It has been suggested that the antitumor properties are associated with intercalation of the anthracycline ring of the antibiotic into the DNA of rapidly proliferating neoplastic cells and subsequent blocking of RNA synthesis (72-75).
[5]

The crystal structures of daunomycin (76) and its analogs (77) have been solved recently to atomic resolution. Aromatic rings B, C and D of the anthracycline ring are planar with ring A in a half-chair conformation and its associated acetyl functional group in an equatorial position. A hydrogen bond between O-7 and OH-9 limits the flexibility about the glycosidic bonds relating the anthracycline and sugar rings (78).

Several molecular models of the daunomycin·DNA complex have been proposed based on fiber diffraction X-ray patterns (79) and physico-chemical measurements (80,81). The models emphasize intercalation of the planar protion of the anthracycline ring between unwound base pairs and electrostatic interactions between the NH$_3^+$ group on the sugar ring and the phosphate group on the backbone (74,75,79,81). The models differ in the details of the overlap geometry between the anthracycline ring and adjacent base pairs at the intercalation site, as to whether the sugar residue is in the minor or major groove, as well as which phosphate is involved in the electrostatic interactions (75,81).

Daunomycin and adriamycin bind to DNA with similar binding constants of $\sim 2.5 \times 10^6$ M^{-1} in 0.1 M buffer, pH 7.0 at 20°C, with one anthracycline antibiotic bound per six nucleotides at maximum intercalative binding (80,81). The binding constant is dependent on the salt concentration and decreases by a factor of ~ 5 on proceeding from 0.1 M to 1.0 M NaCl solution (82).

We have attempted to investigate the daunomycin complex with poly(dA-dT) in order to set constraints on possible overlap geometries in the intercalation complex (26) using methods described in the previous section on the proflavine complex (25). There are nonexchangeable proton markers on ring D and exchangeable proton markers on ring B of the planar portion of the

anthracycline ring. It should be possible to monitor these
groups as well as the nonexchangeable glycosidic proton at H-1'
on the sugar ring. There have been suggestions that daunomycin
exhibits a sequence specificity in its complex with DNA (75) and
we intend to probe this feature using ^{31}P NMR spectroscopy. The
spectral studies were undertaken in high salt solution (1 M NaCl)
to insure intercalation of the antitumor agent between base pairs
and to minimize outside binding along the grooves of the DNA.

Nucleic Acid Exchangeable Protons: We have recorded the
exchangeable proton NMR spectra of the daunomycin•poly(dA-dT)
complex in 1 M NaCl solution between 10 and 16 ppm in an attempt
to monitor the thymidine H-3 proton in the antibiotic-nucleic acid
complex. The studies were undertaken on the Nuc/D = 11.8 and 5.9
complexes with typical spectra presented in Figure 25. The
stronger resonances in the spectra correspond to the thymidine
H-3 proton (Figure 25) and the chemical shifts in the complex are
listed in Table VI. The data demonstrate that the base pairs are
intact in the complex and that the thymidine H-3 proton of
poly(dA-dT) shifts 0.15 ppm to higher field on formation of the
Nuc/D = 5.9 daunomycin complex (Table VI). This suggests that
the ring current contributions from the anthracycline ring more
than compensate that from a base pair displaced as a result of
intercalation. It follows that the thymidine H-3 protons must be
located in the ring current shielding region of the anthracycline
aromatic chromophore at the intercalation site.

Antibiotic Exchangeable Protons: Two exchangeable resonances
(designated by asterisks, Figure 25) are observed between 11.5 and
12.3 ppm in the daunomycin•poly(dA-dT) proton NMR spectra in 1 M
NaCl solution. These resonances correspond to the protons on the
daunomycin ring since their area relative to the thymidine H-3
resonance increases with increasing daunomycin concentration in
the complex (Figure 25). The only candidates for the two exchangeable protons resonating as far downfield as 12 ppm would be the
aromatic hydroxyl protons on ring B. Intramolecular hydroxyl-
carbonyl hydrogen bonds between positions 11-12 and 6-5 should
stabilize thse hydroxyl groups against exchange, as would the
shielding of these hydroxyl groups from solvent as a result of
sandwiching the anthracycline groups between base pairs in the
intercalation complex.

These phenolic hydroxyls have been observed in the proton NMR
spectrum of N-acetyldaunomycin in chloroform solution (83) though
they are broadened out due to rapid exchange with solvent in the
spectrum of daunomycin in H_2O solution. We can therefore compare
the H-6 and H-11 ring B hydroxyl chemical shifts of 12.3 and 11.5
ppm for the Nuc/D = 5.9 daunomycin•poly(dA-dT) complex in aqueous
solution (Figure 25) with the values of 13.86 and 13.15 ppm
observed for N-acetyldaunomycin in nonpolar solution (83).

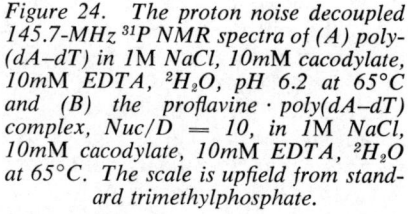

Figure 24. The proton noise decoupled 145.7-MHz ^{31}P NMR spectra of (A) poly-(dA–dT) in 1M NaCl, 10mM cacodylate, 10mM EDTA, ^2H_2O, pH 6.2 at 65°C and (B) the proflavine · poly(dA–dT) complex, Nuc/D = 10, in 1M NaCl, 10mM cacodylate, 10mM EDTA, ^2H_2O at 65°C. The scale is upfield from standard trimethylphosphate.

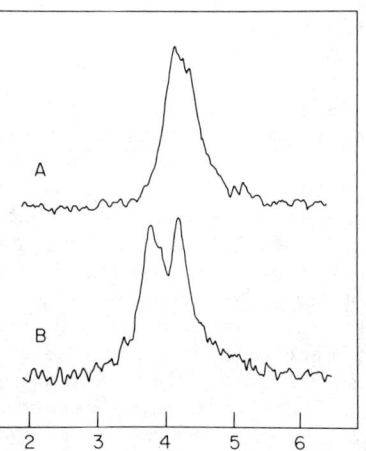

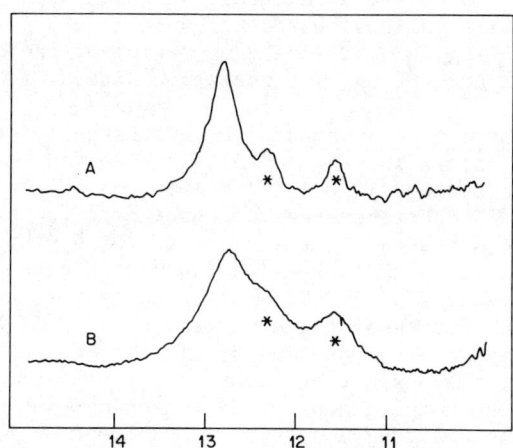

Figure 25. The 360-MHz correlation proton NMR spectra of the daunomycin · poly(dA–dT) complex in 1M NaCl, 10mM cacodylate, 10mM EDTA, 80% H_2O–20% ^2H_2O. Spectrum A corresponds to the Nuc/D = 11.8 complex, pH 6.0 at 67°C and Spectrum B corresponds to the Nuc/D = 5.9 complex, pH 6.05 at 57°C. The strong resonance corresponds to thymidine H-3 proton of the nucleic acid while the weaker resonances (designated by asterisks) corresponds to hydroxyl protons at Positions 9 and 11 on Ring B of the anthracycline ring of daunomycin in the complex.

TABLE VI

Chemical Shift of the Thymidine H-3 Proton
in Poly(dA-dT) and its Daunomycin Complex
in 1 M NaCl Solution

	Chemical Shift, ppm		
	Poly(dA-dT)[a]	Daunomycin·Poly(dA-dT)[b]	
		Nuc/D = 11.8	Nuc/D = 5.9
57°C	12.91	12.86	12.74
67°C	12.84	12.77	12.70

[a] Poly(dA-dT) in 1 M NaCl, 10 mM cacodylate, 0.1 mM EDTA, H_2O, pH 6.53.

[b] Daunomycin·poly(dA-dT) complex in 1 M NaCl, 10 mM cacodylate, 10 mM EDTA, H_2O. Nuc/D = 11.8 complex at pH 6.0 and Nuc/D = 5.9 complex at pH 6.05.

TABLE VII

Chemical Shifts of Anthracycline Ring B
OH Groups at Positions 6 and 11 in the
Daunomycin·Poly(dA-dT) Complex[a]

	Chemical Shift, ppm	
	Nuc/D = 11.8 Complex[b]	Nuc/D = 5.9 Complex[c]
57°C	---, 11.51	12.32, 11.575
67°C	12.275, 11.52	---, 11.40

[a] Buffer was 1 M NaCl, 10 mM cacodylate, 10 mM EDTA, H_2O.

[b] pH = 6.0.

[c] pH = 6.05.

The large upfield shifts of ∼1.6 ppm for both phenolic protons which are located between rings B and C of the anthracycline ring chromophore demonstrates that the anthracycline ring intercalates into the nucleic acid duplex and experiences ring current contributions from adjacent base pairs. This in turn requires that anthracycline rings B and C overlap with adjacent base pairs at the intercalation site.

Nonexchangeable Proton Spectra: Proton spectra of the daunomycin·poly(dA-dT) complex in 1 M NaCl, $^{2}H_2O$ solution have been recorded as a function of temperature at Nuc/D ratios of 50, 25, 9 and 5. Typical spectra (4.0 to 9.0 ppm) for the Nuc/D = 25 complex between 69 and 88°C are presented in Figure 26. The stronger resonances correspond to the base and sugar protons in the complex (in the presence of excess nucleic acid) and these shift as average peaks through the melting transition with increasing temperature (Figure 26). The anthracycline aromatic ring D protons (7.2 to 7.7 ppm) and the anomeric sugar H-1' proton (5.1 to 5.5 ppm) of daunomycin (designated by asterisks in Figure 26) are broadened beyond detection in the duplex state (<75°C) of the Nuc/D = 25 complex but can be followed above 80°C, where they shift as average resonances during the dissociation of the complex (Figure 26).

The temperature dependence of the adenosine H-2 resonance for poly(dA-dT) and the Nuc/D = 50, 25, 9 and 5 complexes in 1 M NaCl are plotted in Figure 27. The adenosine H-2 resonance shifts as an average peak in all the daunomycin complexes, demonstrating fast exchange on the NMR time scale. Biphasic melting curves are observed at the intermediate Nuc/D ratios (Figure 27) and the transition midpoints for antibiotic-free base pair regions and those centered about bound daunomycin are summarized in Table VIII. Daunomycin stabilizes the synthetic DNA duplex since the melting transition midpoint increases from 72.6°C for poly(dA-dT) to 89.6°C in the Nuc/D = 5 antibiotic complex in 1 M NaCl solution.

The melting transition of the daunomycin·poly(dA-dT) complex can also be monitored at the nucleic acid resonance line widths and the data for the adenosine H-8 resonance are plotted in Figure 28. The resonance is very broad at temperatures below the melting transition of the complexes (dashed curves in Figure 28) indicative of stiffening of the synthetic DNA by the bound anthracycline ring.

Base Proton Complexation Shifts: The complexation shifts of certain nucleic acid base resonances of poly(dA-dT) on formation of the daunomycin neighbor exclusion complex reflect the shielding contribution due to the anthracycline ring less the contribution from one neighboring base pair which is displaced following intercalation. Thus, the adenosine H-2 resonance remains unperturbed (Figure 27) while the thymidine exchangeable H-3 proton

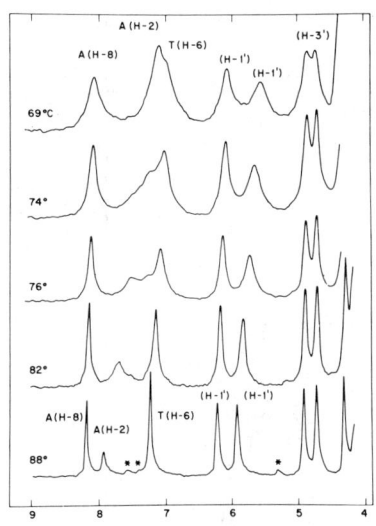

Figure 26. The temperature dependence of the 360-MHz proton NMR spectra (4.5 to 9.0 ppm) of the daunomycin · poly(dA–dT) complex, Nuc/D =25, in 1M NaCl, 10mM cacodylate, 1mM EDTA, ^2H_2O, pH 6.5 between 69° and 88°C. The daunomycin resonances are designated by an asterisk.

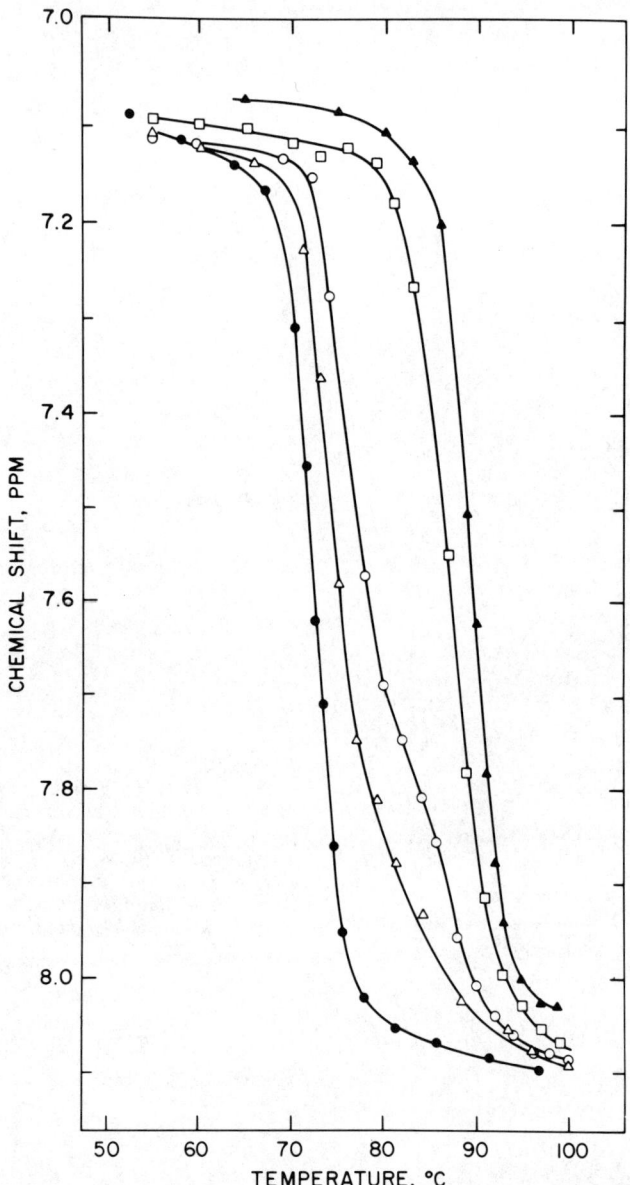

Figure 27. The temperature dependence of the adenosine H-2 resonance (7.1 to 8.1 ppm) for poly(dA–dT) (●) and the daunomycin · poly(dA–dT) complexes, Nuc/D = 50 (△), 25 (○), 9 (□), and 5 (▲) in 1M NaCl, 10mM cacodylate, 1mM EDTA, ^2H_2O solution. The poly(dA–dT) concentration was fixed at 19.3mM in phosphates and the daunomycin concentration was varied to make the different Nuc/D ratio complexes.

TABLE VIII

The Transition Midpoints of the
Daunomycin·Poly(dA-dT) Complexes[a]

	Transition Midpoint, °C		Daunomycin
	Nucleic Acid[b]		
	Antibiotic-Free	Antibiotic-Bound	
No drug	72.6°C	--	
Nuc/D = 50	74.3°	∿85.5°C	
= 25	75.9°	86.8°	88.5 ± 2°C
= 9	82.6°	87.5°	
= 5	--	89.6°	

[a]Buffer was 1 M NaCl, 10 mM cacodylate, 1 mM EDTA, ^2H_2O solution.

[b]The nucleic acid transition midpoints are based on the adenosine H-2 resonance except when Nuc/D = 9, where it is based on the thymidine CH_3-5 resonance.

in the center of the duplex shifts upfield by 0.15 ppm (Table VI, Figure 25) and the thymidine CH_3-5 which is directed towards the major groove shifts upfield by 0.1 ppm (Figure 29). It should be noted that such an upfield shift of the thymidine CH_3-5 group was not observed in the intercalation complexes of ethidium (11), proflavine (25), terpyridylplatinum II (11) and nitroaniline dication with poly(dA-dT). The results require that at least one thymidine CH_3-5 group project onto the periphery of the anthracycline ring system at the intercalation site.

Anthracycline Proton Complexation Shifts: The aromatic protons on ring D of the anthracycline ring have been monitored during the dissociation of the daunomycin·poly(dA-dT) complex at Nuc/D = 25, 9 and 5 in 1 M NaCl solution and these data are plotted between 7.2-7.7 ppm in Figure 30. These ring D protons undergo small upfield shifts on daunomycin complex formation (Table IX, Figure 30) in contrast to the large upfield shifts at the mutagen protons (Table V, Figure 19) observed on proflavine complex formation. This suggests that daunomycin ring D protons project onto the periphery of the nucleic acid base shielding contours and that anthracycline ring D not overlap with nearest-neighbor base pairs at the intercalation site.

Sugar Proton Complexation Shifts: There are small chemical shift changes ($\leq$0.05 ppm) at the nucleic acid sugar H-1' protons on formation of the daunomycin·poly(dA-dT) complex in 1 M NaCl solution (Figure 29). These shifts probably reflect small changes in the glycosidic torsion angles associated with the generation of the intercalative binding site.

The anomeric proton at position 1' on the sugar ring undergoes an $\sim$0.25 ppm upfield shift on daunomycin complex formation in contrast to the CH_3 group at position 5', which undergoes a negligible complexation shift (Figure 30). It remains unclear whether this sugar H-1' complexation shift reflects changes in the glycosidic torsion angles linking the anthracycline and sugar rings or reflects shielding effects of nucleic acid functional groups on this proton in the complex.

Binding Groove: We have monitored the 260 nm absorbance melting curves for synthetic DNA's containing bulky halogen atoms at the pyrimidine 5 position and their Nuc/D = 5 daunomycin complexes in 10 mM buffer solution. The increases in transition midpoints on formation are summarized in Table X. The pyrimidine 5 position projects into the major groove of the double helix. There is additional stabilization in the daunomycin complexes with the halogen substituted DNA's (Table X) which suggests that either daunomycin binds to the minor groove or that binding occurs in the major groove without unfavorable steric contacts between the antibiotic and bulky atoms at position 5 of the pyrimidine ring.

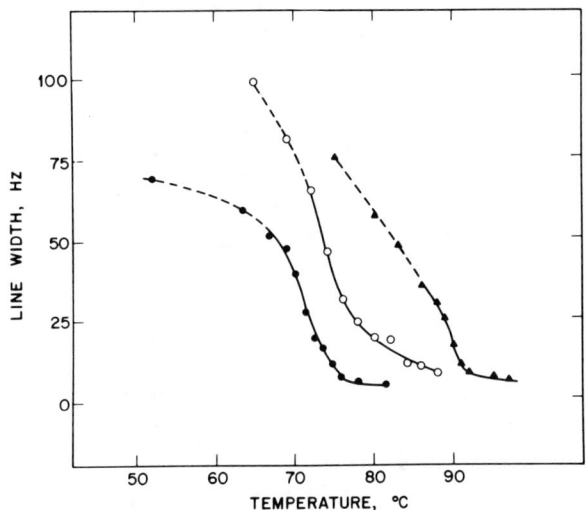

Figure 28. The temperature dependence of the linewidths of the adenosine H-8 resonance for poly(dA–dT) (●) and the daunomycin · poly(dA–dT) complexes, Nuc/D = 25 (○) and Nuc/D = 5 (▲) in 1M NaCl, 10mM cacodylate, 1mM EDTA, ^2H_2O solution. The melting and premelting region are represented by (———) and (– – –), respectively.

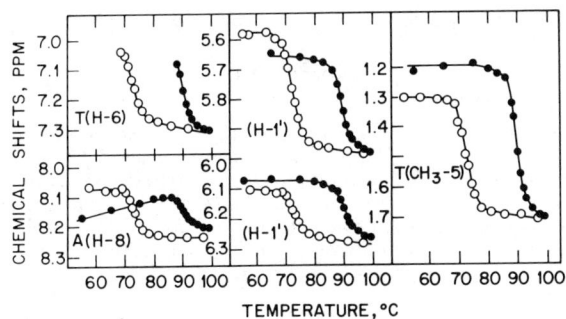

Figure 29. The temperature dependence of the base and sugar H-1' resonances for poly(dA–dT) (○) and the daunomycin · poly(dA–dT) complex, Nuc/D = 5 (●) in 1M NaCl, 10mM cacodylate, 1mM EDTA, ^2H_2O solution

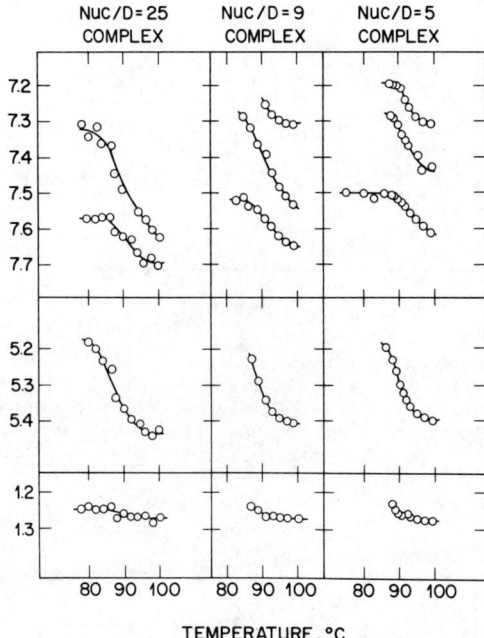

Figure 30. The temperature dependence of the daunomycin anthracycline Ring D aromatic protons (7.3 to 7.7 ppm), the anomeric sugar H-1' proton (5.1 to 5.5 ppm), and the anomeric CH_3-5' proton (1.2 to 1.3 ppm) of the daunomycin · poly-(dA–dT) complex, Nuc/D = 25, 9 and 5 in 1M NaCl, 10mM cacodylate, 1mM EDTA, ^2H_2O. The poly(dA–dT) concentration was fixed at 19.3mM in phosphates and the daunomycin concentration was varied to make the different Nuc/D ratio complexes.

TABLE IX

Experimental Upfield Anthracycline
Complexation Shifts on Formation of The
Daunomycin·Poly(dA-dT) Complex

Assignment[a]	Experimental Shifts, ppm		
	Complex[b]	Free[c]	$\Delta\delta$
H-1/3	7.330	7.680	0.350
H-2	7.570	7.755	0.185

[a] The anthracycline ring D protons are either doublets (H-1, H-3) or a quartet (H-2). We are currently unable to differentiate between H-1 and H-3 for the two doublets.

[b] Data for the daunomycin·poly(dA-dT) complex, Nuc/D = 25 in 1 M NaCl, 10 mM cacodylate, 1 mM EDTA, ^2H_2O, pH 6.5, at 80°C.

[c] Chemical shift of 2 mM daunomycin in 0.1 M phosphate, 1 mM EDTA, ^2H_2O at 90°C.

[d] The experimental upfield complexation shift is the difference between the values for the intact complex at 80°C relative to free daunomycin at 90°C.

Sequence Specificity: We have monitored the 477 nm absorbance of 0.02 mM daunomycin on gradual addition of sequence specific tetranucleotides in 0.1 M phosphate at 1°C. The binding curves are similar for dC-dC-dG-dG and dG-dG-dC-dC at this temperature (Figure 31) and hence daunomycin does not appear to discriminate between pyrimidine(3'-5')purine and purine(3'-5')pyrimidine binding sites at the tetranucleotide duplex level in solution.

Phosphodiester Linkages: The proton noise decoupled ^{31}P NMR spectra of the daunomycin·poly(dA-dT) complex in 1 M NaCl solution at 67°C have been recorded at 1 antibiotic per ∿6 base pairs (Nuc/D = 11.8) and 1 antibiotic per ∿3 base pairs (Nuc/D = 5.9). Resolved resonances are observed for the complex at both Nuc/D ratios (Figure 32). One of the resonances in the complex exhibits a chemical shift similar to that observed for poly(dA-dT) in 1 M NaCl alone (∿4.1 ppm) at this temperature while the other resonance is shifted downfield by 0.3 ppm in the Nuc/D = 11.8 complex and by 0.45 ppm in the NucD = 5.8 complex (Table XI). The results suggest that daunomycin intercalates at either the dTpdA or dApdT sites, resulting in a downfield shift of the ^{31}P resonance of the corresponding phosphodiester grouping at the intercalation site.

Thus, though sequence specificity was not observed in daunomycin complexes at the tetranucleotide duplex level (Figure 31), it appears that the antitumor antibiotic differentiates between pyrimidine(3'-5')purine and purine(3'-5')pyrimidine sites in alternating purine-pyrmidine synthetic DNA's in solution (Figure 32).

Overlap Geometry at the Intercalation Site: We shall attempt to utilize the nucleic acid base and anthracycline ring proton complexation shifts to deduce which anthracycline aromatic ring(s) overlap with nearest neighbor base pairs in the daunomycin·poly-(dA-dT) intercalation complex. It should be noted that the nonplanarity of ring A in the antibiotic requires that the aromatic portion of the anthracycline chromophore cannot intercalate with its long axis colinear to the direction of the Watson-Crick hydrogen bonds at the intercalation site as was demonstrated for proflavine-nucleic acid complexes.

We have demonstrated that anthracycline ring D does not overlap with adjacent base pairs based on the chemical shift of the nonexchangeable protons of this ring system in the complex (Figure 30). The hydroxyl protons at positions 6 and 11 on ring B are intramolecularly hydrogen-bonded to their adjacent carbonyl groups on ring C and hence serve as proton markers for rings B and C. Since these exchangeable protons shift ∿1.6 ppm upfield on complex formation (Figure 25), this portion of the anthracycline ring system must overlap with adjacent base pairs at the intercalation site.

The X-ray structure of a complex of 2 daunomycins per dC-dG-dT-dA-dC-dG duplex is near completion in Professor Alexander

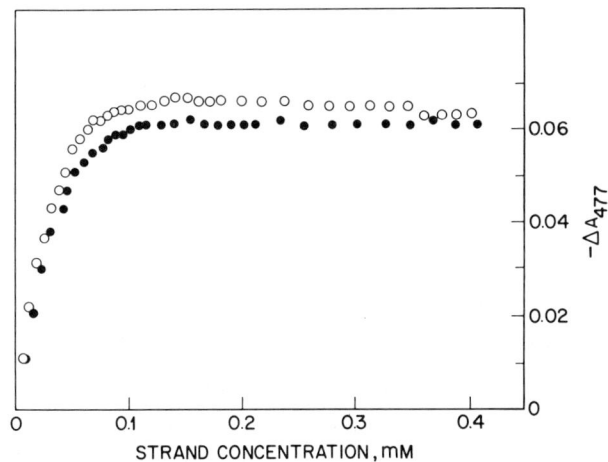

Figure 31. A plot of the change in the 477-nm absorbance of 0.02mM daunomycin on addition of dC–dC–dG–dG (○) and dG–dG–dC–dC (●) in 0.1M phosphate solution, pH 7.0, 1°C

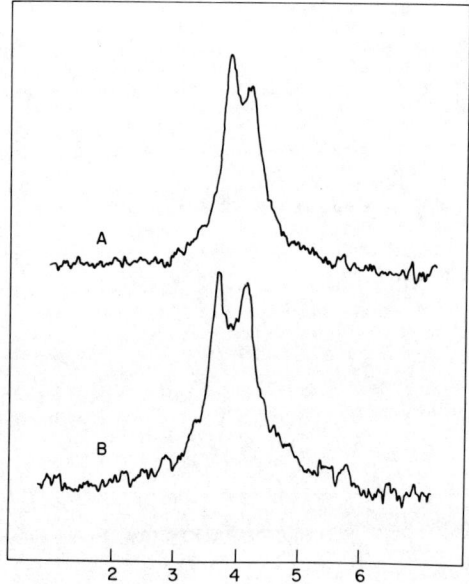

Figure 32. The proton noise decoupled 145.7-MHz ^{31}P NMR spectra of the daunomycin · poly(dA–dT) complex in 1M NaCl, 10mM cacodylate, 10mM EDTA, 80% H_2O–20% D_2O. Spectrum A corresponds to the Nuc/D = 11.8 complex, pH 6.0 at 67°C and Spectrum B corresponds to the Nuc/D = 5.9 complex, pH 6.05 at 67°C. The chemical shifts are upfield from standard trimethylphosphate.

TABLE X

Duplex to Strand Transition Midpoints ($t_{\frac{1}{2}}$, °C) Monitored at the 260 nm Absorbance Band of Daunomycin·Synthetic DNA Complexes in Solution[a]

	Synthetic DNA $t_{\frac{1}{2}}$, °C	Nuc/D=5 Complex $t_{\frac{1}{2}}$, °C	$\Delta t_{\frac{1}{2}}$, °C
Poly(dA-dU)	39.5°	68.5°	29.0°
Poly(dA-br^{5}dU)	50.5°	86.0°	35.5°
Poly(dA-i^{5}dU)	48.5°	86.5°	38.0°
Poly(dI-dC)	33.5°	66.0°	32.5°
Poly(dI-br^{5}dC)	44.5°	81.5°	37.0°

[a] Buffer is 10 mM cacodylate, 1 mM EDTA, H$_2$O, H 6.85. Synthetic DNA concentration was 0.15 mM.

TABLE XI

^{31}P Chemical Shifts of Resolved Doublets in Daunomycin·Poly(dA-dT) Complex at 67°C[a]

Complex	^{31}P Chemical Shift, ppm[b]	
Nuc/D = 11.8	3.88	4.185
Nuc/D = 5.9	3.66	4.11

[a] 1.0 M NaCl, 10 mM cacodylate, 10 mM EDTA, 80% H$_2$O - 20% D$_2$O, pH 6.0.

[b] The chemical shifts are upfield from internal standard trimethylphosphate.

Rich's laboratory. We were informed that the sugar residue
resides in the minor groove and that the intercalation occurred
at dC-dG sites. We recently received the overlap geometry at the
intercalation site observed in the daunomycin·hexanucleotide
crystal (kindly provided by Professor Rich prior to publication)
and it is schematically shown below. [6]
   The positions of the nonexchangeable protons on ring D are
designated by (o), those of the exchangeable protons on ring B by
(•), the thymidine H-3 protons by (*) and the thymidine CH_3 groups
by (x). The NMR data on the daunomycin·poly(dA-dT) complex are in
agreement with the overlap geometry observed in the daunomycin·
hexanucleotide crystal. Thus, the NMR parameters require that
anthracycline rings B and C overlap with adjacent base pairs (as
monitored at exchangeable protons 6 and 11, Figure 25) while
anthracycline ring D project right through the intercalation site
(as monitored at nonexchangeable protons 1, 2, 3 and OCH_3 at 4,
Figure 30 and Table IX) as observed in the crystal. The thymidine
H-3 imino proton of poly(dA-dT) shifts upfield on addition of
daunomycin (Table VI) consistent with intact base pairs in the
complex and location of these Watson-Crick protons over the
anthracycline aromatic ring system at the interaction site. The
upfield shift of the thymidine CH_3 on complex formation (Figure
29) requires that the anthracycline ring system be directed
towards one of the pyrimidine 5 positions, a feature that is observed in the overlap geometry in the crystalline complex. In
addition, the observed binding through the minor groove in the
crystal is consistent with investigations on the binding of
daunomycin to synthetic DNA's with bulky halogen substituents in
the major groove in solution (Table X). Finally, the pyrimidine
(3'-5')purine specificity observed in the crystalline daunomycin·
hexanucleotide complex finds a parallel in the observation of
resolved ^{31}P NMR spectra for the daunomycin·poly(dA-dT) complex
(Figure 32).

Summary: NMR spectroscopy has provided insights into structural
aspects of the daunomycin·poly(dA-dT) complex in solution. These
studies have demonstrated that the anthracycline ring intercalates between intact base pairs with the sugar residues most
likely residing in the minor groove. The complexation shifts of
the anthracycline aromatic exchangeable and nonexchangeable
protons require that rings B and C but not ring D overlap with
adjacent base pairs at the intercalation site. The nucleic base
proton complexation shifts restrict possible overlap orientations
between the thymidine CH_3 groups and the anthracycline ring while
the sugar H-1' proton complexation shifts suggest glycosidic torsion angle change(s) on formation of the intercalation site.
   The observation of partially resolved ^{31}P resonances in the
daunomycin·poly(dA-dT) complex requires that the antibiotic
exhibit a preferred specificity for either the pyrimidine(3'-5')
purine or purine(3'-5')pyrimidine binding sites.

The conclusions on the daunomycin·poly(dA-dT) complex in solution are in good agreement with the detailed X-ray structure at atomic resolution of the daunomycin·dC-dG-dT-dA-dC-dG complex in the crystalline state.

Netropsin-DNA Complexes

The basic oligopeptide netropsin, isolated from streptomycis netropsis exhibits antifungal, antibacterial and antiviral activities and inhibits DNA and RNA tumor viruses in mammalian cells (85-88). The antibiotic binds to double-stranded DNA and stabilizes its thermal melting transition, while no stabilization is observed for single-stranded DNA, double-stranded RNA or RNA·DNA hybrids. Netropsin complexes to dA·dT or dI·dC base pair regions but not to dG·dC base pair regions, while halogen atoms at the pyrimidine 5 position have no effect on the binding (85-88). Binding studies demonstrate that netropsin spans ∿3 base pairs of the nucleic acid double helix and that the complex is stabilized by hydrogen bonding and electrostatic interactions. These data suggest that netropsin binds in the minor groove of dA·dT rich base pair regions of double helical DNA (85-88).[7]

Netropsin complexes with nucleic acids have been monitored by spectroscopic techniques at the oligomer duplex and DNA level (89-96). Our research focussed on the application of high resolution NMR spectroscopy to elucidate structural and dynamic aspects of netropsin complexes with the self-complementary octanucleotide dG-dG-dA-dA-dT-dT-dC-dC duplex (97) and the synthetic DNA poly(dA-dT) (27) in aqueous solution.

Hydrogen Bonding: The thymidine H-3 proton is readily observed in netropsin·poly(dA-dT) complexes in 0.1 M NaCl solution with typical spectra recorded at 47°C for the Nuc/D = 25.5 and 11.8 complexes presented in Figure 33. The spectral region 10 to 16 ppm contains a single exchangeable resonance (Figure 33) which shifts somewhat to lower field with increasing netropsin concentration (Table XII). The Watson-Crick imino hydrogen bonds are therefore intact in the netropsin·poly(dA-dT) complex. The downfield shift at the imino thymidine H-3 proton on interaction of poly(dA-dT) with netropsin (Table XII) may be contrasted with the upfield complexation shifts observed with proflavine (Figure 16) and daunomycin (Table VI) complexes with the synthetic DNA.

Biphasic Absorbance Melting Transition: The melting transition of the netropsin·poly(dA-dT) complex has been monitored as a function of Nuc/D ratio in 0.1 M buffer solution. The transition for the synthetic DNA and Nuc/D = 10 complex give monophasic melting curves with distinct biphasic melting transitions observed at Nuc/D = 50, 25 and 18 ratios (Figure 34). The two cooperative transitions differ by 19°C and are assigned to the opening of antibiotic-free and antibiotic-bound base pair regions in the

[5]

[6]

[7]

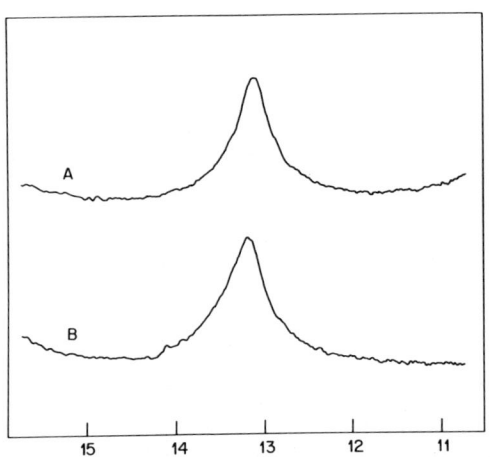

Figure 33. The 360-MHz correlation proton NMR spectra of the netropsin · poly(dA–dT) complex in 0.1M NaCl, 10mM cacodylate, 10mM EDTA, 80% H_2O–20% ^2H_2O. Spectrum A corresponds to the Nuc/D = 25.5 complex, pH 6.1 at 47°C while Spectrum B corresponds to the Nuc/D = 12.8 complex, pH 6.15 at 47°C.

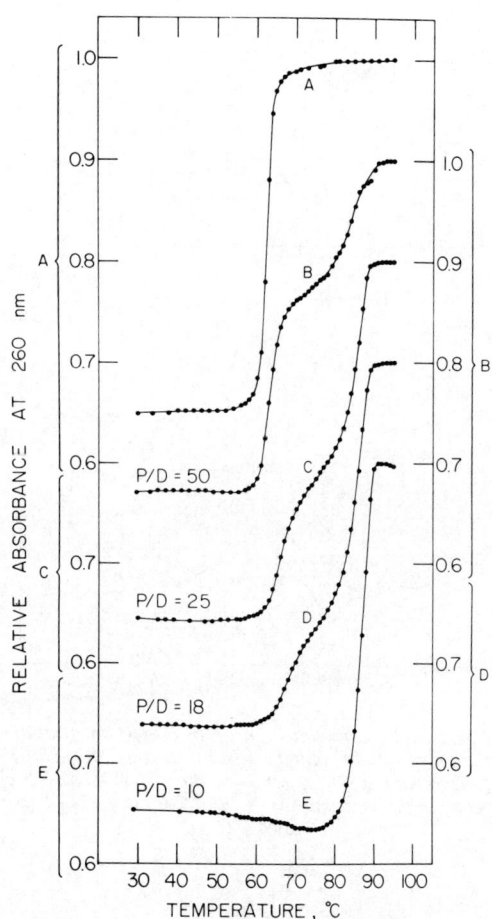

Figure 34. The 260-nm absorbance melting curves (first heating) of poly(dA–dT) and its netropsin complex, Nuc/D = 50, 25, 18, and 10 in 0.1M cacodylate, 4.4mM EDTA, H_2O, pH 7.4. The poly(dA–dT) concentration was fixed at 0.545mM in phosphates and the curves are normalized to an absorbance of 1.0 at 95°C.

netropsin•poly(dA-dT) complex (Table XIII). Biphasic melting transitions have also been reported for the netropsin•poly(dA-dT) complex in low ionic strength (98). By contrast, monophasic melting curves were observed for netropsin•DNA complexes in 0.02 M salt solutions (99). This probably reflects the opening of antibiotic-free base pair regions at hairpin loops and branch points in the netropsin•poly(dA-dT) complex. These types of initiation sites are absent in netropsin•DNA complexes.

Lower Temperature Cooperative Transition: The NMR parameters for the dissociation of the Nuc/D = 50 netropsin•poly(dA-dT) complex have been carefully investigated in 0.1 M phosphate solution, conditions under which 1 netropsin is bound for every 25 base pairs. The nonexchangeable proton NMR spectra (5 to 9 ppm) of the lower temperature cooperative transition between 53.2°C and 68.5°C are presented in Figure 35.

The base and sugar H-1' resonances in the Nuc/D = 50 netropsin•synthetic DNA complex are partially resolved at 53.2°C. These resonances decrease in area on increasing the temperature with the narrower resonances, corresponding to the strand state, increasing in area between 53°C and 69°C (Figure 35). The low temperature cooperative melting transition of the netropsin•poly(dA-dT) complex corresponds to the opening of antibiotic-free base pair regions and hence it follows that this transition is slow on the NMR time scale in the complex of 1 netropsin per 25 base pairs. This is strikingly demonstrated by the doubling of the peaks in the 61°C spectrum of the complex (Figure 35), which corresponds to the midpoint of the lower temperature cooperative transition. These results demonstrate that the duplex dissociation rate constants of the antibiotic-free base pair regions in the Nuc/D = 50 netropsin•poly(dA-dT) complex (Figure 35) are at least an order of magnitude smaller than their corresponding values in poly(dA-dT)(Table III).

Antibiotic-Induced Conformational Changes: The temperature dependent chemical shifts of the adenosine and thymidine sugar H-1' protons in poly(dA-dT) and the Nuc/D = 50 netropsin•poly-(dA-dT) complex in 0.1 M phosphate are plotted between 40° and 100°C in Figure 36. It is striking that even though only 1 netropsin is bound per 25 base pairs, the resonances corresponding to antibiotic-free and antibiotic-bound base pair regions have shifted to high field to the same extent (Figure 36). Thus, the average chemical shift of the upfield sugar H-1' resonance (5.47 ppm) in the Nuc/D = 50 netropsin•poly(dA-dT) complex at 53.2°C is the same as the chemical shift of this resonance (designated by an asterisk) for base pairs centered about bound netropsin in the partially melted out complex at 68.5°C (Figure 35). This demonstrates that netropsin alters the nucleic acid conformation at its binding site and that the structural perturbation is propagated to adjacent antibiotic-free base pair regions (27). A

TABLE XII

Chemical Shift of the Thymidine H-3 Proton
in Poly(dA-dT) and Its Netropsin Complex

	Poly(dA-dT)	Chemical Shift, ppm Netropsin·Poly(dA-dT) Complex[b]	
		Nuc/D = 25.5	Nuc/D = 12.8
47°C	12.89	13.08	13.20

[a] Poly(dA-dT) in 0.1 M Phosphate, 1 mM EDTA, H_2O.

[b] Netropsin·poly(dA-dT) complex in 0.1 M NaCl, 10 mM cacodylate, 10 mM EDTA, H_2O. Nuc/D = 25.5 complex at pH 6.1 and Nuc/D = 12.8 complex at pH 6.15.

TABLE XIII

Biphasic Duplex to Strand Transition Midpoints
($t_{1/2}$, °C) monitored at the 260 nm Absorbance
of Netropsin·Poly(dA-dT) Complexes in Solution[a]

	$t_{1/2}$, °C	
Nuc/D[b]	Antibiotic-Free Base Pair Regions	Antibiotic-Bound Base Pair Regions
Poly(dA-dT) alone	62.5	--
50	63.5	84.5
25	66.0	85.5
18	67.5	86.0
10	--	87.0

[a] Buffer is 0.1 M cacodylate, H_2O, pH 7.4.

[b] Synthetic DNA concentration was fixed at 0.545 mM in nucleotides.

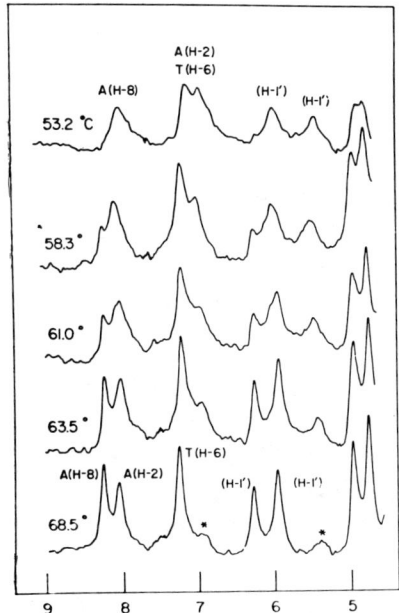

Figure 35. The temperature dependence (first heating) of the 360-MHz proton NMR spectra (5 to 9 ppm) of the netropsin · poly(dA–dT) complex, Nuc/D = 50, in 0.1 cacodylate, 4.4mM EDTA, ^2H_2O, pH 7.5 between 53° and 68.5°C. The asterisks designate some of the minor resonances from base pairs centered at the binding site.

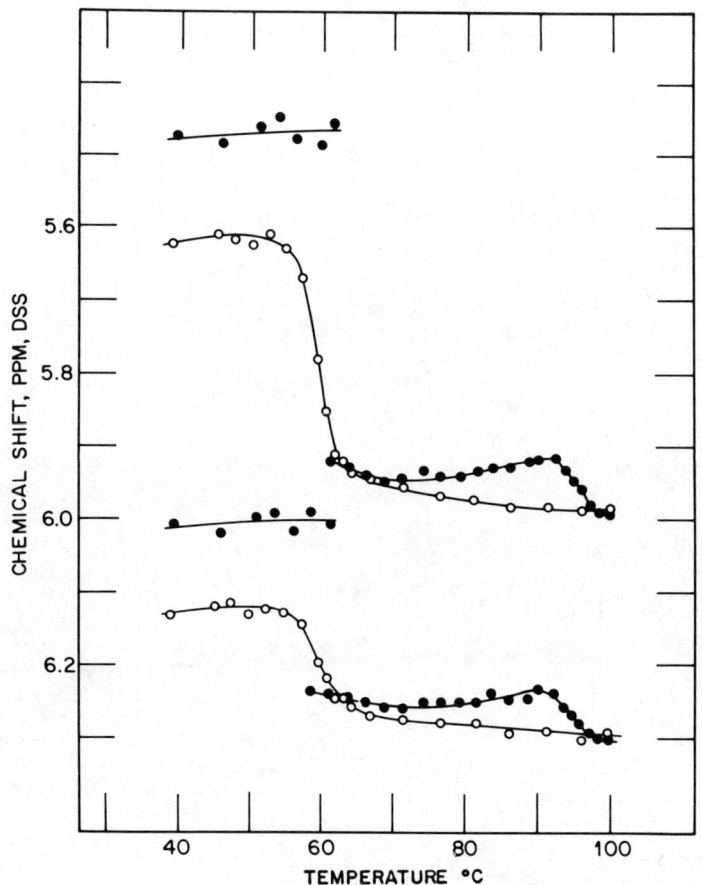

Figure 36. The temperature dependence (40° to 100°C) of the sugar H-1' chemical shifts of poly(dA–dT) (○) and the netropsin · poly(dA–dT) complex, Nuc/D = 50 (●) in 0.1M cacodylate, 4.4mM EDTA, ^2H_2O solution

similar result has been reported recently for complexes of the related antibiotic distamycin with calf thymus DNA (100).

The sugar H-1' protons undergo selective upfield shifts on interactions of netropsin with poly(dA-dT), in contrast to the base protons, which are perturbed to a much smaller degree on complex formation (Figure 37). These results demonstrate that the antibiotic induced conformational changes primarily occur at the adenosine and thymidine glycosidic torsion angles with less pronounced perturbations in the base pair overlap geometries.

Netropsin Migration Along Partially Opened DNA: The plots of the sugar H-1' chemical shifts in the Nuc/D = 50 netropsin·poly(dA-dT) complex in 0.1 M buffer demonstrate that the lower temperature and higher temperature cooperative transitions exhibit midpoints of 61°C and 95°C, respectively (Figure 36). This section covers the NMR spectral parameters for the complex between these temperature values (65 and 90°C) with typical spectra in the 5 to 9 ppm presented in Figure 38.

The NMR spectrum of the Nuc/D = 50 complex at 66°C (Figure 35) contains major resonances corresponding to base and sugar protons of antibiotic-free base pair regions that have melted out and minor resonances (represented by asterisks) which corresponds to base pair regions centered about bound antibiotic. The observation of separate resonances suggests that the migration of netropsin along the partially opened duplex is slow on the NMR time scale at 66°C. The minor resonances exhibit a constant area between 64° and 71°C and the relative area of the minor and major resonances require that ∿20 base pairs are associated with the opening of antibiotic-free base pair regions when 1 netropsin binds for every 25 base pairs.

The base proton chemical shifts and line widths for poly(dA-dT) and for the major resonances in the Nuc/D = 50 netropsin·poly(dA-dT) complex between 70° and 100°C are plotted in Figure 39. The line widths of the major resonances corresponding to the antibiotic-free strand regions in the 1 netropsin per 25 base pair complex are much larger than the corresponding values for poly(dA-dT) in the strand state between 70°C and 80°C (Figure 39). This demonstrates a decreased segmental mobility for antibiotic strand regions anchored at one or both ends by netropsin-bound base pair regions in the netropsin·poly(dA-dT) complex between 65° and 85°C.

The minor resonances corresponding to base pair regions centered about bound netropsin broaden and coalesce with the major resonances corresponding to melted out antibiotic-free base pair regions when the temperature is raised from 70° to 85°C for the Nuc/D = 50 netropsin·poly(dA-dT) complex (Figure 38). There is also a small upfield shift (Figures 36 and 39) and a small increase in line width (Figure 39) at the base and sugar H-1' resonances associated with increasing temperature from 70° to 90°C. This suggests that netropsin migration along the partially

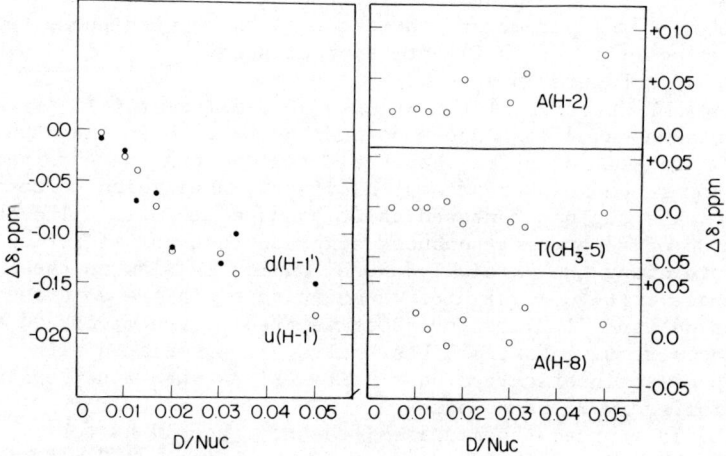

Figure 37. The variation of the sugar H-1' and base proton (adenosine H-8, H-2, and thymidine CH_3-5) chemical shifts of 24mM poly(dA–dT) in 0.1M cacodylate, 10mM EDTA, ^2H_2O, pH 7.25, 53°C on gradual addition of netropsin

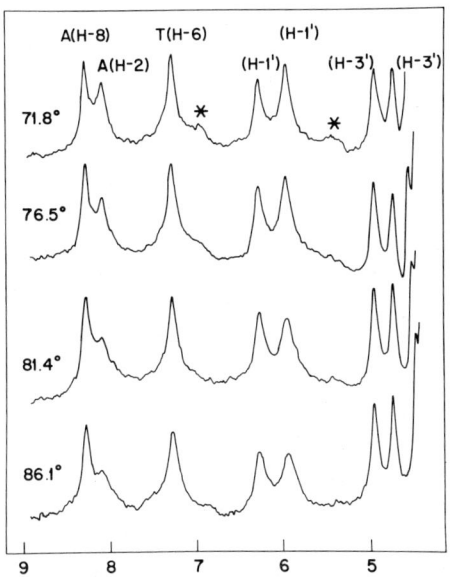

Figure 38. The temperature dependence (first heating) of the 360-MHz proton NMR spectra (5 to 9 ppm) of the netropsin · poly(dA–dT) complex, Nuc/D = 50 in 0.1M cacodylate, 4.4mM EDTA, ^2H_2O, pH 7.25 between 71° and 85°C. The asterisks designate some of the minor resonances from base pairs centered at the binding site.

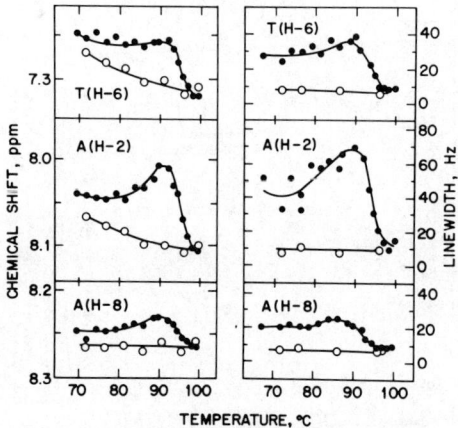

Figure 39. The temperature dependence of the chemical shifts and linewidths of the base resonances of poly(dA–dT) (○) and the Nuc/D = 50 netropsin · poly-(dA–dT) complex (●) in 0.1M cacodylate, 4.4mM EDTA, ^2H_2O, pH 7.25 between 70° and 100°C

opened duplex in the Nuc/D = 50 complex becomes fast on the NMR time scale above 75°C such that the NMR spectrum corresponds to an average of antibiotic-free and -bound regions at this high temperature (Figure 38).

Higher Temperature Cooperative Transition: The dissociation of base pair regions centered about bound netropsin in the Nuc/D = 50 complex in 0.1 M buffer can be monitored at the nucleic acid proton chemical shifts (Figures 36 and 39) and line widths (Figure 39) between 85° and 100°C. The proton resonances shift downfield as average peaks (Figures 36 and 39) and decrease in line widths (Figure 39) to their corresponding values in poly-(dA-dT) with increasing temperature. The transition midpoint occurs at ∼95°C. The dissociation rate constant for the Nuc/D = 50 netropsin·poly(dA-dT) complex at 95°C is fast compared to the 0.1 ppm chemical shift difference for the adenosine H-2 resonance between 90°C and 100°C (Figure 39).

Phosphodiester Groups: The ^{31}P NMR studies (previous section) of the proflavine·poly(dA-dT) complex (Figure 24), the daunomycin·poly(dA-dT) complex (Figure 32, Table XI), and a nitrophenyl reporter molecule·poly(dA-dT) complex demonstrated that the dTpdA and dApdT phosphodiester linkages gave partially resolved phosphorus spectra. These two resonances in the intercalation complexes mentioned above were separated by about 0.4 ppm and centered about 4.0 ppm upfield from the trimethylphosphate reference. One of these ^{31}P resonances in the intercalation complex exhibits approximately the same chemical shift as in poly(dA-dT) while the other, corresponding to the phosphodiester at the intercalation site, shifts downfield on complex formation. Downfield shifts of this magnitude ($\leq$0.4 ppm) appear to be characteristic of the intercalation site though much larger downfield shifts (2.0 ± 0.5 ppm) have been reported by our group for the actinomycin D·nucleic acid complexes (64,65).

In striking contrast, we have observed upfield shifts in the netropsin·poly(dA-dT) complex in 0.1 M buffer with ^{31}P resonances at 4.25 and 4.56 ppm (upfield from standard trimethylphosphate) in the Nuc/D = 12.8 complex at 47°C (Figure 40). A similar upfield ^{31}P complexation shift has been reported for the netropsin·octanucleotide complex (97) and a tetralysine·octanucleotide complex (101). The peptide antibiotic and the oligopeptide bind in the grooves of the nucleic acid and the upfield ^{31}P shifts may reflect either the contributions of electrostatic interactions and/or a conformational change in the nucleic acid phosphodiester groups on complex formation.

Antibiotic Resonances: The protons on the N-methylpyrrole rings of netropsin can be detected in the 1 antibiotic per 5 base pair netropsin·poly(dA-dT) complex at 87°C (Figure 41). The proton line widths indicate that the entire complex is intact at this

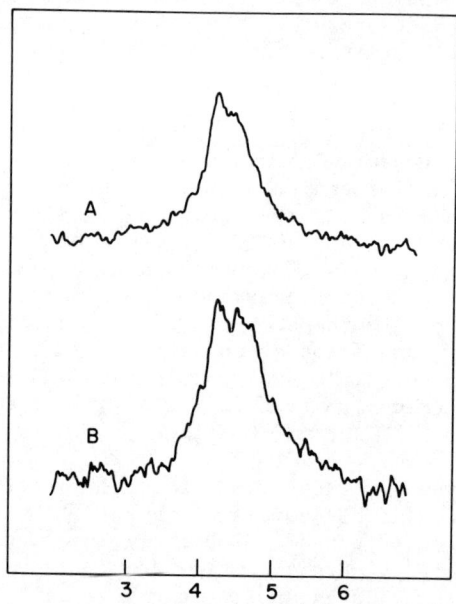

Figure 40. The proton noise decoupled 145.7-MHz ^{31}P NMR spectra of the netropsin · poly(dA–dT) complex in 0.1M NaCl, 10mM cacodylate, 10mM EDTA, 80% H_2O–20% ^2H_2O. Spectrum A corresponds to the Nuc/D = 25.5 complex, pH 6.1 at 47°C while Spectrum B corresponds to the Nuc/D = 12.8 complex, pH 6.15 at 47°C.

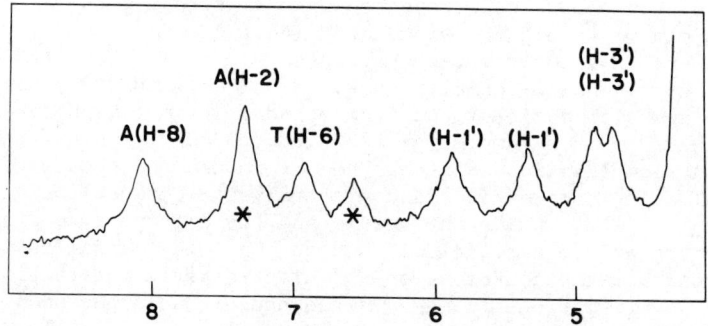

Figure 41. The 360-MHz proton NMR spectrum (4.5 to 9.0 ppm) of the Nuc/D = 10 netropsin · poly(dA–dT) complex in 0.1M cacodylate, 4.4mM EDTA ^2H_2O, pH 7.08 at 87°C. The tentatively assigned N-methylpyrrole protons of netropsin are designated by asterisks.

this temperature with the two pyrrole H-5 protons superimposed at 6.58 ppm (designated by asterisks) and the adenosine H-2 nucleic acid and the two pyrrole H-3 protons (designated by asterisks) superimposed at 7.35 ppm (Figure 41). A comparison of these pyrrole proton complexation shifts with the corresponding values in free netropsin indicates that complex formation results in downfield shifts at the H-3 position and upfield shifts at the H-5 position (Table XIV). The observed downfield shifts at the pyrrole H-3 protons suggest that they experience in-plane ring current contributions of the base pairs following binding of the netropsin in the groove of DNA.

The exchangeable and nonexchangeable netropsin protons have been monitored in detail for the 1 antibiotic per dG-dG-dA-dA-dT-dT-dC-dC octanucleotide duplex complex in our laboratory (97). That study has proved insights into the role of hydrophobic, hydrogen bonding and electrostatic interactions to the stability of the peptide antibiotic-nucleic acid complex in solution (97).

Summary: High resolution proton NMR spectroscopy has been utilized to probe the opening of antibiotic-free base pair regions ($t_{1/2} \sim 61°C$) and those centered about bound netropsin ($t_{1/2} \sim 90°C$) in the biphasic dissociation of the 1 netropsin per 25 base pair poly(dA-dT) complex in 0.1 M buffer solution with temperature.

The observation of selective complexation shifts in the nucleic acid resonances of the synthetic DNA demonstrate a change in the glycosidic torsion angles of the adenosine and thymidine residues and a minimal perturbation in the base pair overlaps on addition of netropsin. These structural perturbations at the antibiotic binding site are propagated to adjacent antibiotic-free base pair regions at low netropsin concentrations.

We observe that netropsin binds tightly to DNA and stabilizes ∼5 base pairs centered about its binding site. The opening rates of the intervening base pair stretches during the dissociation of the antibiotic-free base pair regions in the Nuc/D = 50 complex are slower by an order of magnitude compared to the dissociation rates for the duplex to strand transition of poly(dA-dT) alone in 0.1 M buffer solution.

The NMR data demonstrate a decreased segmental mobility of antibiotic-free strand regions anchored at one or both ends by netropsin-bound base pair regions in the partially melted out Nuc/D = 50 complex between 65° and 90°C. The migration of the antibiotic along the partially opened structure is slow between 60° and 70°C but the migration rate increases between 75° and 90°C.

Conclusion

This review outlines our approach to investigating the structure and dynamics of synthetic DNA's and their complexes with mutagens and antibiotics by high resolution NMR spectrosocpy in

TABLE XIV

Experimental Antibiotic Pyrrole Proton
Complexation Shifts ($\Delta\delta$, ppm) in the
Netropsin·Poly(dA-dT) Complex[a]

Pyrrole Protons	Free Netropsin[b]	Netropsin·Poly(dA-dT) Complex, Nuc/D = 10[c]	Complexation Shift, $\Delta\delta$[d]
H-3[e]	7.13, 7.07	(7.35)[f]	+0.22, +0.28
H-5[e]	6.79, 6.73	6.58	-0.21, -0.15

[a] Buffer is 0.1 M cacodylate, 4.4 mM EDTA, ^2H_2O, pH 7.08.

[b] Chemical shift at 96.8°C corresponding to dissociated components in complex.

[c] Chemical shift at 87.0° corresponding to intact complex.

[d] The experimental complexation shift represents the difference between the Nuc/D = 10 netropsin·poly(dA-dT) complex at 87.0°C and the dissociated compoents at 96.8°C.

[e] There are protons at positions H-3 and H-5 on each of the two pyrrole rings of netropsin. The set at higher field in free netropsin is tentatively assigned to H-5 while the set at lower field is tentatively assigned to H-3.

[f] The assignment of the pyrrole H-3 protons in the complex is not definitive since they cannot be monitored as average resonances during the dissociation of the complex.

solution. We have focussed our efforts on alternating purine-pyrimidine polynucleotides and demonstrated that the inherent symmetry and high degree of branching in these synthetic DNA's results in well resolved proton resonances exhibiting moderate widths in the duplex state which shift as average peaks through the melting transition. The proton markers are distributed throughout the base pairs and the sugar residues and monitor aspects of the hydrogen bonding, base stacking and torsion angle changes in the sugar-phosphate backbone during the non-cooperative premelting transition and cooperative melting transition.

The effect of counterion on the synthetic DNA conformation has been probed by comparing the NMR parameters in high salt sodium and tetramethylammonium solutions. We observe a specific change at one of the two glycosidic torsion angles and at one of the two phosphodiester linkages for poly(dA-dT) on proceeding from Na^+ to TMA^+ solutions.

These NMR studies have been extended to probe the interaction of ligands with synthetic DNA's to deduce conformational information on the binding site in aqueous solution. The proflavine·poly(dA-dT) complex can be independently monitored at the mutagen and nucleic acid proton resonances during the dissociation of the complex as a function of temperature. The NMR parameters of proflavine on complex formation readily demonstrate intercalation of the mutagen between intact base pairs with the long axis of the proflavine ring colinear with the direction of the Watson-Crick hydrogen bonds.

Structural information on the daunomycin·poly(dA-dT) complex was only obtained after monitoring both the exchangeable phenolic and nonexchangeable benzylic protons on the aromatic part of the anthracycline ring system. The complexation shifts demonstrate that anthracycline rings B and C overlap with adjacent base pairs while no such overlap is observed for anthracycline ring D which passes through the intercalation site. The observed complexation shift at the thymidine CH_3-5 resonance requires that the long axis of the anthracycline ring be approximately normal to the direction of the Watson-Crick hydrogen bonds.

The purine(3'-5')pyrimidine and the pyrimidine(3'-5')purine phosphodiester linkages are partially resolved in the proflavine and daunomycin intercalation complexes with poly(dA-dT) with the phosphodiester at the intercalation site shifting to low field. This suggests that these intercalating agents exhibit a sequence specificity in their complexes with alternating purine-pyrimidine polynucleotides.

The NMR parameters for the complex of the groove binding antibiotic netropsin with poly(dA-dT) are quite distinct from those observed with the intercalating proflavine and daunomycin ligands with the synthetic DNA in aqueous solution. Netropsin induces a conformational change at its binding site which is propagated to adjacent antibiotic-free base pair regions at low netropsin concentrations. This structural change occurs primarily at the glycosidic torsion angles of the purine and pyrimidine residues with minimal perturbation at the base pair overlap geometries. We have been able to monitor the opening of antibiotic-free base pair regions and those centered about bound netropsin during the biphasic dissociation of the complex, as well as the migration of netropsin along the partially melted out complex as a function of temperature.

The proflavine, daunomycin and netropsin complexes with DNA have been investigated by crystallographic and spectroscopic techniques in solution. The NMR results presented above contribute to and extend the knowledge of drug-DNA complexes gained by various physico-chemical measurements. We shall attempt in the future to extend our NMR investigations to the study of additional classes of DNA binding antibiotics of current importance as antitumor agents used in combination chemotheraphy. The NMR methodology outlined above is also amenable to investigations of protein-nucleic acid interactions in solution.

References

1. Patel, D. J. (1974), in Peptides, Polypeptides and Proteins, eds. Blout, E. R., Bovey, F. A., Goodman, M. and Lotan, N. J., Wiley and Sons, New York, 459-471.
2. Kallenbach, N. R. and Berman, H. M., (1977), Quarterly Reviews Biophysics, 10, 138-236.
3. Kearns, D. R., (1977), Ann. Revs. Biophys. Bioeng., 6, 477-523.
4. Krugh, T. R. and Nuss, M. E., (1979), in Biological Applications of Magnetic Resonance, ed., Shulman, R. G., Academic Press, New York 113-175.
5. Kearns, D. R. and Shulman, R. G., (1974), Acc. Chem. Res., 7, 33-39.
6. Kearns, D. R., (1976), Prog. Nucleic Acids Research, 18, 92-149.
7. Reid, B. R. and Hurd, R. E., (1977), Acc. Chem. Res., 10, 396-402.
8. Patel, D. J., (1978), Ann. Revs. Phys. Chem., 29, 337-362.
9. Robillard, G. T., and Reid, B. R., (1979), in Biological Applications of Magnetic Resonance, ed. Shulman, R. G., Academic Press, New York, 45-112.
10. Patel, D. J. (1979), Acc. Chem. Res., 12, 118-125.
11. Patel, D. J. (1980), in Nucleic Acid Geometry and Dynamics, ed., Sarma, R., Pergamon Press, New York, 185-231.
12. Barkeley, M. S. and Zimm, B. H., (1979), J. Chem. Phys., 70, 2991-3008.
13. Bolton, P. H. and James, T. L., (1979), J. Am. Chem. Soc., 102, 25-31.
14. Early, T. A. and Kearns, D. R., (1979), Proc. Natl. Acad. Scs. U.S.A., 76, 4165-4169.

15. Hogan, M. E. and Jardetzky, O., (1979), Proc. Natl. Acad. Scs. U.S.A., 76, 6341-6345.
16. Baldwin, R. L., (1971), Acc. Chem. Res., 4. 265-272.
17. Baldwin, R. L., (1968), in Molecular Associations in Biology, Academic Press, New York, 145-162.
18. Patel, D. J. and Canuel, L. L. (1976), Proc. Natl. Aca. Scs. U.S.A., 73, 674-678.
19. Gale, E. F., Cundliffe, E., Reynolds, P. E., Richmond, M. H., and Waring, M. J., (1972), Molecular Basis of Antibiotic Action, 173-278.
20. Meienhofer, J. and Atherton, E., (1977), in Structure-Activity Relationships Among the Semi-Synthetic Antibiotics, Academic Press, New York, 427-529.
21. Neidle, S., (1979), Prog. Medicinal Chemistry, 16, 151-221.
22. Sobell, H. M., Tsai, C. C., Jain, S. C., and Gilbert, S. G., (1977), J. Mol. Biol., 114, 333-365.
23. Gabbay, E., (1977), in Bioorganic Chemistry, Vol. III, ed. vanTamelen, E. E., Academic Press, New York, 33-70.
24. Patel, D. J. (1978), J. Polym. Science: Polym. Symposium, 62, 117-141.
25. Patel, D. J., (1977), Biopolymers, 16, 2739-2754.
26. Patel, D. J., (1978), Eur. J. Biochem., 90, 247-254.
27. Patel, D. J. and Canuel, L. L., (1977), Proc. Natl. Acad. Scs. U.S.A., 74, 5207-5211.
28. Kearns, D. R., Patel, D. J., and Shulman, R. G., (1971), Nature, 229, 338-340.
29. Crothers, D. M., Hilbers, C. W., and Shulman, R. G., (1973), Proc. Natl. Acad. Scs. U.S.A., 70, 2899-2901.
30. Patel, D. J. and Tonelli, A. E., (1974), Biopolymers, 13, 1943-1964.
31. Kallenbach, N. R., Daniel, W. E. Jr., and Kaminker, M. A., (1976), Biochemistry, 15, 1218-1224.
32. Hilbers, C. W., (1979), in Biological Applications of Magnetic Resonance, ed., Shulman, R. G., Academic Press, New York, 1-44.
33. Giessner-Prettre, C., Pullman, B., Borer, P. N., Kan, L. S. and T'so, P. O. P., (1976), Biopolymers, 15, 2277-2286.
34. Arter, D. B., and Schmidt, P. G., (1976), Nucleic Acids Research, 3, 1437-1447.
35. Early, T. A., Kearns, D. R., Burd, J. F., Larson, J. E. and Wells, R. D., (1977), Biochemistry, 16, 541-551.
36. Crothers, D. M., (1964), J. Mol. Biol., 9, 712-733.
37. Palecek, D., (1976), Progr. Nucl. Acid. Res. Mol. Biol., 18, 151-213.
38. Brahms, S., Brahms, J., and VanHolde, K. E., (1976), Proc. Natl. Aca. Scs. U.S.A., 73, 3453-3457.
39. Sobell, H. M., (1980), in Nucleic Acid Geometry and Dynamics, ed., Sarma, R., Pergamon Press, 289-323.
40. Spatz, H. C., and Baldwin, R. L., (1965), J. Mol. Biol., 11, 213-222.

41. Shindo, H., Simpson, R. T., and Cohen, J. S., (1979), J. Biol. Chem., 254, 8125-8128.
42. Shindo, H., and Zimmerman, S. B., (1979), Nature, 76, 2703-2707.
43. Shapiro, J. T., Stannard, B. S., and Felsenfeld, G., (1969), Biochemistry, 8, 3233-3241.
44. Patel, D. J. and Canuel, L. L. (1979), Proc. Natl. Acad. Scs. U.S.A., 76, 24-28.
45. Lerman, L. S., (1961), J. Mol. Biol., 3, 18-30.
46. Lerman, L. S., (1963), Proc. Natl. Acad. Scs. U.S.A., 49, 94-102.
47. Waring, M. J., (1970), J. Mol. Biol., 54, 247-279.
48. Berman, H. M., Neidle, S., Stallings, W., Taylor, G., Carrell, H. L., Glusker, J. P., and Achari, A., (1979), Biopolymers, 18, 2405-2429.
49. Reddy, B. S., Seshadri, T. P., Sakore, T. D., and Sobell, H. M., (1979), J. Mol. Biol., 135, 787-812.
50. Shieh, H-S., Berman, H. M., Dabrow, M. and Neidle, S., (1980), Nucleic Acids Research, 8, 85-97.
51. Alden, C. J. and Arnott, S., (1975), Nucleic Acids Research, 2, 1701-1717.
52. Alden, C. J., and Arnott, S., (1977), Nucleic Acids Research, 4, 3855-3861.
53. Giessner-Prettre, C. and Pullman, B., (1976), C. R. Acad. Sc. Paris, 283D, 675-677.
54. Giessner-Prettre, C. and Pullman, B., (1977), J. Theor. Biol., 65, 171-188.
55. Wang, A. H. J., Nathans, J., van der Marel, G., Van Boom, J. H. and Rich, A., (1978), Nature, 276, 471-474.
56. Giessner-Prettre, C. and Pullman, B., (1976), Biochem. Biophys. Res. Commun., 70, 578-581.
57. Schara, R. and Muller, W., (1972), Eur. J. Biochem., 29, 210-216.
58. Krugh, T. R., (1972), Proc. Natl. Acad. Scs. U.S.A., 69, 1911-1914.
59. Patel, D. J., (1976), Biochem. Biophys. Acta., 442, 98-108.
60. Davanloo, P. and Crothers, D. M., (1976), Biochemistry, 15, 4433-4438.
61. Krugh, T. R., Wittlin, F. M. and Cramer, S. P., (1975), Biochemistry, 14, 197-210.
62. Tsai, C. C., Jain, S. C. and Sobell, H. M. (1975), Phil. Trans. Roy. Soc., London, B272, 137-146.
63. Davanloo, P. and Crothers, D. M., (1976), Biochemistry, 15, 5299-5305.
64. Patel, D. J., (1974), Biochemistry, 13, 2396-2402.
65. Patel, D. J., (1976), Biopolymers, 15, 533-558.
66. Patel, D. J. and Canuel, L. L., (1976), Proc. Natl. Acad. Scs. U.S.A., 73, 3343-3347.
67. Patel, D. J. and Canuel, L. L., (1977), Proc. Natl. Acad. Scs. U.S.A., 74, 2624-2628.

68. Kastrup, R. V., Young, M. A. and Krugh, T. R., (1978), Biochemistry, 17, 4855-4865.
69. Chiao, Y. C., Gurudath Rao, K., Hook, J. W., Krugh, T. R. and Sengupta, S. K., (1979), Biopolymers, 18, 1749-1762.
70. Patel, D. J. and Canuel, L. L., (1979), Eur. J. Biochem., 96, 267-276.
71. Gueron, M. and Shulman, R. G., (1975), Proc. Natl. Acad. Scs. U.S.A., 72, 3482-3485.
72. DiMarco, A., Arcamone, F. and Zunino, F., (1974), in Antibiotics III, eds. Corcoran, J. and Hahn, F. E., Springer-Verlag, New York, 101-128.
73. Brown, J. R., (1978), Prog. Med. Chem., 15, 125-164.
74. Henry, D. W., (1976), in Cancer Chemotheraphy, ed., Sartorelli, A. C. American Chemical Society, Washington, 15-57.
75. Neidle, S., (1978), in Topics in Antibiotic Chemistry II, ed., Sammes, P. G., Ellis Harwood Ltd., Chichester, 241-277.
76. Neidle, S. and Taylor, G., (1977), Biochem. Biophys. Acta., 479, 450-459.
77. VonDreele, R. B. and Einck, J. J., (1977), Acta. Crystallogr. B., 33, 3283-3288.
78. Neidle, S. and Taylor, G. L., (1979), FEBS Letters, 107, 348-354.
79. Pigram, W. J., Fuller, W., and Hamilton, L. D., (1972), Nature New Biol., 235, 17-19.
80. Zunino, F., Gambetta, R., DiMarco, A., Luoni, G., and Zaccara, A., (1976), Biochem. Biophys. Res. Commun., 69, 744-750.
81. Gabbay, E. J., Grier, D., Fingerle, R. E., Reimer, R., Levy, R., Pearce, S. W., and Wilson, W. D., (1976), Biochemistry, 15, 2062-2070.
82. Zunino, F. Gambetta, R., DiMarco, A., Velcich, A., Zaccara, A., Quadrifoglio, F., and Crescenzi, V., (1977), Biochem. Biophys. Acta., 476, 38-46.
83. Arcamone, F., Cassinelli, G., Franceshi, G., Orezzi, P., and Mondelli, R., (1968), Tetrahedron Letters, 30, 3353-3356.
84. Arcamone, F. Cassinelli, G., DiMatteo, F., Forenza, S., Ripamonti, M. C., Rivola, G., Vigerani, A., Clardy, J., and McCabe, T., (1980), J. Am. Chem. Soc., 102, 1462-1463.
85. Zimmer, Ch., (1975), Progr. Nucleic Acid Research Molecular Biology, 15, 285-318.
86. Hahn, F. E., (1974), Antibiotics, III, eds. Corcoran, J. W. and Hahn, F. E., Springer-Verlag, New York, 285-318.
87. Gursky, G. V., Tumanyan, V. G., Zasedatelev, A. S., Zhuze, A. L., Grokhovsky, S. L., and Gottikh, B. P., (1977), Nucleic Acid-Protein Recognition, ed. Vogel, H. J., Academic Press, New York, 189-196.
88. Wartell, R. M., Larson, J. E. and Wells, R. D., (1974), J. Biol Chem., 249, 6719-6731.

89. Berman, H. M., Neidle, S., Zimmer, S., and Thrum, H., (1979), Biochemica et. Biophysica Acta., 561, 124-130.
90. Gurskaya, G. V., Grokhovsky, S. L., Zhuze, A. L., and Gottikh, B. P., (1979), Biochemica et. Biophysica Acta., 563, 336-342.
91. Wartell, R. M., Larson, J. E. and Wells, R. D., (1975), J. Biol. Chem., 250, 2698-2702.
92. Burd, J. F., Wartell, R. M. Dodgson, J. B., and Wells, R. D., (1975), J. Biol. Chem., 250, 5109-5113.
93. Zimmer, Ch., Luck, G. and Fric, I., (1976), Nucleic Acids Research, 3, 1521-1532.
94. Zimmer, Ch., Marck, Ch., Schnider, Ch., and Guschlbauer, W., (1978), Nucleic Acids Research, 6, 2831-2837.
95. Sutherland, J. C., Duval, J. F. and Griffin, K. P., (1978), Biochemistry, 17, 5088-5091.
96. Martin, J. C., Wartell, R. M. and O'Shea, D. C., (1978), Proc. Natl. Acad. Scs. U.S.A., 75, 5483-5487.
97. Patel, D. J., (1979), Eur. J. Biochem., 99, 369-378.
98. McGhee, J. D., (1976), Biopolymers, 15, 1345-1375.
99. Zimmer, Ch., Puschendorf, B., Grunicke, H. Chandra, P. and Venner, H., (1971), Eur. J. Biochem., 21, 269-278.
100. Hogan, M. Dattagupta, N. and Crothers, D. M., (1979), Nature, 278, 521-524.
101. Davanloo, P., Armitage, I. M. and Crothers, D. M., (1979), Biopolymers, 18, 663-680.

RECEIVED July 10, 1980.

INDEX

INDEX

A

Al$_2$O$_3$–liquid interface 6
Actinomycin D on addition of dC–dC–dG–dG and dG–dG–dC–dC, change in absorbance of 252f
Adsorption of PVAc and PS 17
AMP complex, effects of inorganic phosphate and Na$^+$ ion on the enzyme–Mn^{2+}- 59
ATPase(s)
  Gd(II) EPR studies of membrane-bound 49
  Mn(II) EPR studies of membrane-bound 49–80
  –Mn^{2+} complexes with ATP or AMP–PNP 59
  stoichiometry and dissociation constants for Mn^{2+} binding to membrane 52
Adenosine
  H-2 resonance for poly(dA–dT) complexes, temperature dependence of 262f
  H-2 resonances, temperature dependence of the linewidth of the adenosine H-8 and 240f
  H-8 resonance(s)
    and adenosine H-2 resonance, temperature dependence of the linewidth of 240f
    for poly(dA-dT) and the daunomycin · poly(dA–dT) complexes, temperature dependence of the linewidths of 265f
    and thymidine H-3 resonances, temperature dependence of the linewidth of the nonexchangeable 238f
  –uridine sequences in synthetic DNA, alternating 220–235
Adriamycin 256
Alkyl chain motion, pendant 133–143
Alkyl groups in solution, molecular dynamics of 119–145
Alkylammonium ions, counterion binding 235–241
Anthracycline proton complexation shifts for the daunomycin · poly(dA-dT) complex 264

Antibiotic exchangeable protons in the daunomycin · poly(dA–dT) proton NMR spectra 257–260
Antibiotic, netropsin · poly(dA–dT) complex
  -induced conformational changes in 277–281
  pyrrole proton complexation shifts in 288t
  resonances in 285–287
ATP to the ATPase–Gd^{3+} complex, addition of 74
ATPase–Gd^{3+} complex, addition of ATP to 74

B

Backbone motion, polymer 128–133
Base-pair overlaps 226
Base proton complexation shifts of the daunomycin · poly(dA–dT) complex 260–264
Benzene and PE, C-13 spectra of 194
γ-Bidentate CrATP on signal intensity of Mn^{2+}-(Na$^+$ + K$^+$)-ATPase, effect of β 62f
Binding groove 264
Biphasic absorbance melting transition of the netropsin · poly(dA–dT) complex 273–277
Bond cleavage in PVC 40
Branches per polymer molecule, average number of long-chain 111
Branches per ten thousand carbon atoms, long-chain 112
Branching
  ethyl 113
  long-chain
    characteristics of PE 104t
    detection of 111
    PEs, energies of flow activation .. 104t
    in PEs using high-field carbon-13 93–118
    measurements, use of C-13 NMR in .. 96
  short-chain
    determination, C-13 NMR and .. 101
    in high-density PEs, long-chain and 96
    for PEs, long-chain and 115t
Breaking strain, short term 20

Breaking stress(es) 23
and free radicals, relationship
between 25f

C

C-1 carbon
  methylene spin-lattice relaxation
    times for 141t
  NOEFs for 141t
  undergoing internal rotation,
    NOEFs for 123f
C-2 carbon, methylene spin-lattice
  relaxation times for 142t
C-2 carbon, NOEFs for 142t
Ca^{2+}-ATPase 49
  effect of Gd^{3+} on the longitudinal
    water proton relaxation rate
    in solutions containing 68f
  effect of Gd^{3+} on $1/T_1$ of $^7Li^+$ in the
    presence of 67f
  for Gd^{3+}, high affinity of 66
  and Gd^{3+} ion, frequency depend-
    ence of T_{1p} in solutions of 70t
  Gd(III) as a probe in NMR and
    EPR studies of sarcoplasmic
    reticulum 64
  paramagnetic effects of metals with
    Gd^{3+} on 70
  titration of $GdCl_3$ with 69f
  X-band EPR spectrum of Gd^{3+}
    complex with 75f
    in the presence of ATP 75f
$(CHCl_3)$–solid interface 6
1-CP (see 1-Chloropentane)
3-CP (see 3-Chloropentane)
Cab-O-Sil SiO_2 11
Carbon resonances 101
Carbon-13 (C-13)
  chemical shifts from TMS as a
    function of branch length, PE
    backbone and side-chain 100t
  chemical shifts for VCl–VCl_2
    copolymers 85
  long-chain branching in PEs using
    high-field 93–118
  MAS spectra
    of PCTFE 207
      fluorine-decoupled 210f
    of PTFE
      at -72°C 208f
      fluorine-decoupled 208f
      as a function of temperature .. 207
  NMR
    analysis of VCl–VCl_2 copolymers 87t
    in branching measurements,
      use of 96
    for characterizing PEs, advan-
      tages and disadvantages of .. 116

Carbon-13 (C-13) (continued)
  NMR (continued)
    copolymers, spectrum(a)
      of the ethylene-1-octene 103
      from ethylene-1-olefin 97
      of 25.2 MHz of an ethylene-1
        -butene 98f
        -heptene 99f
        -hexene 99f
        -octene 99f
        -pentene 98f
        -propene 98f
    high-resolution 81–91
    molecular structure of VCl–VCl_2
      copolymers by 81–91
    nomenclature 97
    and short-chain branching
      determination 101
    in solid polymers, background
      magic-angle spinning
      for 193–196
    spectrum(a)
      at 25.2 MHz of a low-density
        PE produced from a
        high-pressure process ... 102f
      at 50 MHz of NBS 1483 .... 114f
      of 50 MHz of Polymers
        A, B, C, D, and E ... 105f–109f
      of CPBD 205f
      of PEO at -140°C, proton-
        decoupled/CP 203f
      of semicrystalline PVF_2 .... 210f
    studies of solid polymers,
      VT-MAS 193–217
    NOEF's for neat DD 129t
    NOEF's of PBA in 50% (w/w)
      solution in toluene 126t
    relaxation times for the methyl
      carbons of PP 213f
    spectra
      of benzene and PE 194
      of isotactic PP at 24°C and
        -170°C, proton-decoupled
        CP/MAS 212f
      for polycrystalline PE 195f
      of PP, proton-decoupled
        CP/MAS 211
    spin-lattice relaxation time(s) 119
      for neat DT 131t
      for neat HDD 130t
      of PBA in 50% (w/w) solution
        in toluene 126t
      of PBMA 122t
      of PHMA in 50% (w/w) solu-
        tion in toluene-d_8 .... 125t
Chemical shift
  lineshape changes for the crystalline
    fraction of melt-recrystallized
    PTFE 180

INDEX

Chemical shift (*continued*)
  parameters associated with the melting transition of poly-(dA–dT) in 0.1 M phosphate solution 228*t*
  spectra of PTFE 170–180
1-Chloropentane (1-CP) 38
  ESR spectrum of 38
  at LNT, ESR spectrum of UV irradiated 42*f*
3-Chloropentane (3-CP) 38
  ESR spectrum of 38
  at LNT, ESR spectrum of UV irradiated 39*f*
cis-1,4 Polybutadiene (CPBD) 204–206
  C-13 NMR spectra of 205*f*
  and VT-MAS 204
Cooperative transition for the netropsin · poly(dA–dT) complex, lower-temperature 277
Cooperative transition for the netropsin · poly(dA–dT) complex, higher-temperature 285
Copolymer(s)
  C-13 NMR spectrum of the ethylene-1-octene 103
  C-13 NMR spectra from ethylene-1-olefin 97
  ethylene-1-hexene 110
Counterion binding, alkylammonium ions 235–241
CP/MAS C-13 spectra of PP, proton-decoupled 211
CPBD (*see cis*-1,4 Polybutadiene)
Cross
  polarization (CP) 211
  -relaxation rate for PVF$_2$ 165*f*
  -relaxation rate for 40/60 PVF$_2$/PMMA blend 165*f*

D

Daunomycin
  on addition of dC–dC–dG–dG and dG–dG–dC–dC, change absorbance of 269*f*
  anthracycline ring D aromatic protons, temperature dependence of 266*f*
  –DNA complexes 256–273
  poly(dA–dT) complex(es)
    anthracycline proton complexation shifts for 264
    base proton complexation shifts of 260–264
    intercalation site, overlap geometry at 268–272
    360-MHz correlation proton NMR spectra of 258*f*

Daunomycin (*continued*)
  –DNA complexes (*continued*)
    nonexchangeable proton spectra of 260
    nucleic acid exchangeable protons for 257
    phosphodiester linkages for 268
    proton noise decoupled 145.7-MHz ^{31}P NMR spectra of 270*f*
    sequence specificity for 268
    sugar proton complexation shifts for 264
    temperature dependence
      of the adenosine H-2 resonance for poly(dA–dT) and 262*f*
      of the base and sugar H-1' resonances for poly(dA–dT) and 265*f*
      of the linewidths of the adenosine H-8 resonance for poly(dA–dT) and 265*f*
      of the 360-MHz proton NMR spectra of 261*f*
    transition midpoints of 263*t*
    upfield anthracycline complexation shifts on formation of 267*t*
  · poly(dA–dT) proton NMR spectra, antibiotic exchangeable protons in 257–260
  · synthetic DNA complexes in solution, duplex-to-strand transition midpoints at the 260-nm absorbance band of 271*t*
1,2-Decanediol (DD) 120–124
  C-13 NOEFs for neat 129*t*
  relaxation properties of 124
  spin-lattice relaxation times of neat 127*t*
1,2,3-Decanetriol (DT) 124
  C-13 spin-lattice relaxation times for neat 131*t*
  C-13 T_1's and NOEFs of 136*t*
  NT_1 as a function of carbon number at 22.6 MHz for the alkyl carbons of 137*f*
  NT_1 as a function of carbon number at 67.9 MHz for the alkyl carbons of 137*f*
  relaxation properties of 124
p-Dioxane (DX) 37
Dipolar interaction and line broadening in the ESR spectrum of PVC radicals 44
Distribution widths, spin-lattice relaxation time for 134*f*
DNA
  alternating adenosine–uridine sequences in synthetic 220–235

DNA (*continued*)
  complexes
    daunomycin– 256–273
    netropsin– 273–287
    proflavine– 241–256
  interactions in solution, drug– ..219–294
  netropsin migration along partially
    opened 281–285
  in solution, NMR spectra of 219
Drug–DNA interactions in solution 219–294
DT (*see* 1,2,3-Decanetriol)
Duplex
  dissociation rate constants in 0.1M
    phosphate solution, poly(dA–dT) 248t
  formation, upfield shifts associated with poly(dA–dT) and poly-(dA–dU) 229t
  pair overlaps 226–231
  state for poly(dA–dT), hydrogen bonding in 220–221
  -to-strand transition of poly(dA–dT) 222
  -to-strand transition midpoints at the 260-nm absorbance band of daunomycin · synthetic DNA complexes in solution .. 271t
DX (p-Dioxane) 37
Dynamic range 113

E

Electron paramagnetic resonance (EPR) 50
  Gd(III) 51
    spectrum for, X-band
      in aqueous solution 72f
      complex with Ca^{2+}-ATPase 75f
      in presence of ATP 75f
    properties of, to probe its environment 71
    studies of membrane enzymes, theoretical basis for Mn(II) and 50
  Mn(II) 50–51
    spectra of bound 55–64
      to $(Na^+ + K^+)$-ATPase 56f
    spectra for complexes
      of $(Na^+ + K^+)$-ATPase, K-band
        and ATP 61f
        and β,γ-bidentate CrATP .. 61f
        and $Co(NH_3)_4$ATP 63f
        and α,β,γ-tridentate CrATP 63f
      of $(Na^+ + K^+)$-ATPase and AMP, X-band 58f
      with $(Na^+ + K^+)$-ATPase, X-band 57f

Electron paramagnetic resonance (EPR) (*continued*)
  Mn(II) (*continued*)
    studies of membrane-bound ATPases 49–80
    studies of sheep kidney $(Na^+ + K^+)$-ATPase and Mg^{2+}-ATPase 52
  studies of membrane-bound ATPases, Gd(II) 49
Electron spin resonance (ESR) 19
  investigation of environmental effects on nylon fibers 19–34
  spectrum(a) 38
    of 1-CP
    of 3-CP 38
    of PVAc 3f
      at the solid–liquid interface 7f
    of PVC
      and copolymer glasses 37–38
      radical, computer-simulated .. 45f
      radicals, dipolar interaction and line broadening in 44
      –THF at -110°C 45f
    resulting from mechanical damage of Nylon 6 .... 21f
    on surface coverage of PVAc at the SiO_2-(Cab-O-Sil)-CCl_4 interface, dependence of 10f
    of UV irradiated
      1-CP at LNT 42f
      3-CP at LNT 39f
      PVC–TP glass at LNT 39f
  study of intermediates in the photo-degradation of PVC 35–49
Energy exchange in PVF_2 164
Enzyme
  –Gd^{3+} complex, rotational correlation time of 71
  –Mn^{2+}-AMP complex, effects of inorganic phosphate and Na^+ ion on 59
  system, addition of substrate to 59
EPR (*see* Electron paramagnetic resonance)
ESR (*see* Electron spin resonance)
Ethidium bromide on addition of dC–dC–dG–dG and dG–dG–dC–dC, change in absorbance of 253f
Ethyl branching 113
Ethylene copolymer(s)
  -1-butene, C-13 NMR spectra at 25.2 MHz of 98f
  -1-heptene, C-13 NMR spectra at 25.2 MHz of 99f
  -1-hexene 110
    C-13 NMR spectra at 25.2 MHz of 99f

INDEX

Ethylene copolymers (*continued*)
  -1-octene, C-13 NMR spectrum(a)
    of 103
    at 25.2 MHz of 99f
  -1-olefin, C-13 NMR spectra from 97
  -1-pentene, C-13 NMR spectra at
    25.2 MHz of 98f
  -1-propene, C-13 NMR spectra at
    25.2 MHz of 98f

F

F-19 REV-8 chemical shift
  lineshapes
    of the amorphous fraction of melt
      recrystallized PTFE 178f
    of the crystalline fraction of melt
      recrystallized PTFE 176f
    of a 68% crystalline PTFE
      sample 172f
    of melt-recrystallized PTFE 175f
  spectra of amorphous and crystalline fractions of PTFE, decomposed 173f
  spectra of PTFE samples varying
    crystallinity 171f
FID (*see* Free induction)
Flow activation energies 103
  of long-chain branching of PEs 104t
Fluorine-decoupled, MAS–C-13
  spectra of PCTFE 210f
Fluorine-decoupled, MAS–C-13
  spectra of PTFE 208f
Fluoropolymers and VT-MAS 206–209
Fracture
  free-radical concentration production during 23
  surface of a nylon filament
    exposed to
      an air environment 31f
      an NO_2 environment 28f, 29f
      an O_3 environment 30f
      an SO_2 environment 27f
  surface produced by cutting with
    scissors 29f
Free
  induction (FID) 159
  signal(s)
    illustrating composite character .. 160f
    of plasticized PVC at 75°C 158f
    of pure and plasticized PVC 157f
    subsequent to $T_{1\rho}$ 162f
  -radical concentration production
    during fracture 23
  radicals, relationship between
    breaking stresses and 25f

G

"g" factor 116
^{13}G {^{1}H} NMR spectra of PVC and
  VCl–VCl$_2$ copolymers 82f
Gd^{3+}
  in aqueous solution, X-band EPR
    spectrum for 72f
  to Ca^{2+}-ATPase, binding of 66
  on the Ca^{2+}-ATPase, paramagnetic
    effects of metals with 70
  complex with Ca^{2+}-ATPase, X-band
    EPR spectrum of 75f
    in presence of ATP 75f
  high affinity of Ca^{2+}-ATPase for 66
  on the longitudinal water proton
    relaxation rate in solutions
    containing Ca^{2+}-ATPase,
    effect of 68f
  to probe its environment, EPR
    properties of 71
  on $1/T_1$ of $^7Li^+$ in the presence of
    Ca^{2+}-ATPase, effect of 67f
Gd(III) EPR 51
  studies of membrane enzymes,
    theoretical basis for Mn(II)
    and 50
Gd(III) as a probe in NMR and EPR
  studies of sarcoplasmic reticulum
  Ca^{2+}-ATPase 64–74
$GdCl_3$ with Ca^{2+}-ATPase, titration of 69f
$GdCl_3$ concentration, steady-state
  levels of E–P at 0°C as a function of 65f
Gadolinium II EPR studies of membrane-bound ATPases 49
Gaussian curve, first derivative 44
Gaussian distribution in the amorphous regions of PTFE and a
  TFE/HFP copolymer, temperature dependence on the rms value
  of 187f
Glycosidic torsion angle 237
Goldman–Shen experiment of PVC at
  110°C 164
Goldman–Shen pulse sequence 159, 163t

H

1,2-Hexadecanediol (HDD) 120, 124
  C-13 spin-lattice relaxation times
    for neat 130t
  relaxation properties of 124
Hexamethyl-phosphoric triamide
  (HPMT) 37
Hexamethyldisiloxane (HMDS) 85
High
  -density PE, long- and short-chain
    branching in 96

High (continued)
-field C-13, long-chain branching in PEs using 93–118
resolution C-13 NMR 81–91
Higher-temperature cooperative transition for the netropsin · poly-(dA–dT) complex 285
Hizex 7000, SECs of 95f
Hydrogen bonding 235–237
in duplex state for poly(dA–dT) 220–221
in netropsin · poly(dA–dT) complexes 273

L

Lineshape(s)
changes observed for the amorphous fraction of PTFE 182
changes, rate of the molecular motion responsible for amorphous 183
of PTFE, crystalline and amorphous 170–180
of PTFE, decomposed 174
Liquid nitrogen temperature (LNT) .. 36
Loading frame 21f
Local-order parameters for PTFE 184
Log-χ^2 distribution 128
NOEF for 132f
spin-lattice relaxation time for 132f
Lower-temperature cooperative transition for the netropsin · poly-(dA–dT) complex 277

M

Mn^{2+}
-AMP complex, effects of inorganic phosphate and Na^+ on the enzyme– 59
binding to membrane ATPases, stoichiometry and dissociation constants for 52
EPR spectrum(s)
bound 55–64
to $(Na^+ + K^+)$-ATPase 56f
complexes $(Na^+ + K^+)$-ATPase and AMP, X-band 58f
and ATP, K-band 61f
and β,γ-bidentate CrATP, K-band 61f
and $Co(NH_3)_4ATP$, K-band .. 63f
and α,β,γ-tridentate, CrATP, K-band 63f
X-band 57f
effects of nucleotide substrates on 55
EPR studies
–enzyme complexes, Mn^{2+} 55

Mn^{2+} (continued)
EPR studies (continued)
for measuring metal–substrate distances on $(Na^+ + K^+)$-ATPase 60
of Mn^{2+}–enzyme complexes 55
to Mg^{2+}-ATPase, binding of 54f
to $(Na^+ + K^+)$-ATPase, binding of 53f
–$(Na^+ + K^+)$-ATPase, effect of β,γ-bidentate CrATP on signal intensity of 62f
Mn(II) EPR, studies 50–51
and Gd(III) EPR studies of membrane enzymes, theoretical basis for 50
of membrane-bound ATPases 49–80
of sheep kidney $(Na^+ + K^+)$-ATPase and Mg^{2+}-ATPase 52
Mg^{2+}-ATPase 49
binding of Mn^{2+} 54f
Magic-spinning apparatus 199f
Magnet spectrometers 97
Magic-angle spinning (MAS) 193
apparatus 196–201
assembly 196–201
C-13
NMR in solid polymer, background 193–196
NMR studies of solid polymers, VT 193–217
spectra
of PCTFE, C-13 207
of PCTFE, fluorine-decoupled 210f
of PTFE at -72°C 208f
of PTFE, fluorine-decoupled .. 208f
of PTFE as a function of temperature, C-13 207
techniques for producing high spinning rates for 196
Melt-recrystallized PTFE 170–180
Melting transition
of the netropsin · poly(dA–dT) complex, biphasic absorbance 273–277
of the poly(dA–dT) 237
of proflavine · poly(dA–dT) complexes 242
Methylene spin-lattice relaxation times for C-1 carbon 141t
Methylene spin-lattice relaxation times for the C-2 carbon 142t
MIR (see Multiple internal rotations)
Molecular
dynamics
of alkyl groups in solution 119–145
of polymer chains in solution 119–145
of PTFE 169–191
motion characterization 119–145

INDEX

Molecular *(continued)*
  motion responsible for amorphous
    lineshape changes, rate of 183
Monomer composition, calculation of   84
  VCl–VCl$_2$ copolymers from C-13
    NMR areas 86
Multiple internal rotations (MIR) 120
  C-13 T_1's and NOEFs using the
    theory of 138t
Multiple-pulse NMR of solid
  polymers 169–191
Mutagen resonance 247

N

(Na$^+$ + K$^+$)-ATPase 49
  active site structure of 76f
  binding of 53f
  EPR spectra of Mn^{2+} bound to 56f
  EPR spectra for Mn^{2+} complexes of
    and AMP, X-band 58f
    and ATP, K-band 61f
    and β,γ-bidentate CrATP,
      K-band 61f
    and Co(NH$_3$)$_4$ATP, K-band 63f
    and α,β,γ-tridentate CrATP,
      K-band 63f
    and Mg^{2+}-ATPase, Mn(II) EPR
      studies of sheep kidney 52
  Mn^{2+} EPR studies for measuring
    metal–substrate distances on .. 60
  X-band EPR spectra of Mn^{2+}
    complexes with 57f
NBS 1483, C-13 NMR spectrum
  at 50 MHz of 114f
NBS 1475, SECs of 95f
NO$_2$, degradation in strength of Nylon
  6 as a function of time in 24f
NO$_2$ as a function of time, degradation in strength of Nylon 6 at
  room temperature 25f
Netropsin
  –DNA complexes 273–287
  migration along partially opened
    DNA 281–285
  · poly(dA–dT) complex(es)
    antibiotic
      -induced conformational
        changes in 277–281
    pyrrole proton complexation
      shifts in 288t
    resonances in 285–287
    higher-temperature cooperative
      transition for 285
    hydrogen bonding in 273
    lower-temperature cooperative
      transition for 277
    360-MHz correlation proton
      NMR spectra of 275f

Netropsin *(continued)*
  · poly(dA–dT) complex(es)
    (continued)
    360-MHz proton NMR
      spectrum of 286f
    phosphodiester groups in 285
    proton noise decoupled 145.7-
      MHz ^{31}P NMR spectra of .. 286f
    temperature dependence of the
      360-MHz proton NMR
      spectra of 279f, 283f
    temperature dependence of the
      sugar H-1' chemical shifts
      of poly(dA–dT) and 280f
Nicolet TT-14 pulse-FT spectrometer 201–202
Nitroxide(s) 1–18
  undergoing motional narrowing,
    simulated composite spectra .. 4f
NMR
  relaxation measurements 147
  of solid polymers, multiple-
    phase 169–191
  spectrum(a)
    of the daunomycin · poly-
      (dA–dT) complex, 360-
      MHz correlation proton 258f
    of DNA in solution 219
    360-MHz proton
      of the daunomycin · poly-
        (dA–dT) complex, temperature dependence of .. 261f
      of the netropsin · poly(dA–
        dT) complex 286f
    correlation 275f
    temperature dependence
      of 279f, 283f
    of poly(A-U), temperature
      dependence of 233f
    of poly(dA–dT) 223f, 225f
    of the proflavine · poly-
      (dA–dT) complex 244f
    of the thymidine H-3 proton
      in poly(dA–dT) 221f
    of poly(dA–dT), 360-MHz
      correlation proton 236f
  for studying composite polymeric
    systems, pulsed 147–168
Nuclear Overhauser effects in PEs 112
Nuclear Overhauser enhancement
  (NOE) 119
  factors (NOEFs) 120
    for the C-1 carbon 141t
    undergoing internal rotation .. 123f
    for the C-2 carbon 142t
    for the log-χ^2 distribution 132f
    for neat DD, C-13 .... 129t
    of PBA in 50% (w/w) solution

Nuclear Overhauser enhancement
(NOE) (*continued*)
factors (NOEFs) (*continued*)
in toluene, C-13 .... 126*t*
for PBMA in 50% (w/w) solution in toluene-d_8 .... 122*t*
Nonexchangeable proton spectra of
the daunomycin · poly(dA–dT)
complex 260
Nucleic acid
base resonances 242–247
exchangeable protons for daunomycin · poly(dA–dT) complex 257
sugar resonances 247
Number average molecular weight
data from SEC for PEs 115
Number average molecular weight
for PEs 115*t*
Nylon
fibers, ESR investigation of
environmental effects on 19–34
fibers in NO_2, strength of 23
filament control 27*f*
filament exposed, fracture surface of
to an air environment 31*f*
to an NO_2 environment 28*f*, 29*f*
to an O_3 environment 30*f*
to an SO_2 environment 26*f*, 27*f*
Nylon 6
degradation in strength of, as a
function of time in NO_2 .... 24*f*
at room temperature 25*f*
degradation in strength of,
as a function of time in various
environments 22*f*
effect of ozone on the strength of 20
effects of environment and stress on 19
ESR spectra resulting from
mechanical damage of 21*f*
samples, effect of sustained loading
on the strength of 20

O

1-Olefins 96
Overlap geometry at the daunomycin ·
poly(dA–dT) complex intercalation site 268–272
Overlap geometry for proflavine 251–255
Oxide surfaces, polymers adsorbed on 1–18
Ozone on the strength of Nylon 6,
effect of 20

P

^{31}P NMR spectra, proton noise
decoupled 145.7-MHz
of daunomycin · poly(dA–dT)
complex 270*f*

^{31}P NMR spectra, proton noise decoupled
145.7-MHz (*continued*)
of the netropsin · poly(dA–dT)
complex 286*f*
of poly(dA–dT) 234*f*, 243*f*
and the proflavine · poly(dA–dT)
complex 258*f*
PBA (*see* Poly(*n*-butyl acrylate))
PBMA (*see* Poly(*n*-butyl methacrylate))
PCTFE (*see* Poly(chlorotrifluoroethylene))
PE (*see* Polyethylene)
PEO (*see* Polyethylene oxide)
PHMA (*see* Poly(*n*-hexyl methacrylate))
PS (*see* Polystyrene)
PTFE (*see* Polytetrafluoroethylene)
PVAc (*see* Poly(vinyl acetate))
PVC (*see* Poly(vinyl chloride))
PVF_2 (*see* Poly(vinylidene fluoride))
Pendant alkyl chain motion 133–143
Phillips PE 5003, SECs of 95*f*
Phosphodiester linkage(s) 232, 237–241
for daunomycin · poly(dA–dT)
complex 268
proflavine · poly(dA–dT)
complex 255–256
Photodegradation of PVC, ESR
study of intermediates in 35–49
Photodegradation of PVC, mechanisms of 35–36
Poly(A-U), temperature dependence
of the base and sugar H-1′ proton
chemical shifts of 233*f*
Poly(A-U), temperature dependence
of the 360-MHz proton NMR
spectra of 233*f*
Poly(*n*-butyl acrylate) (PBA) 124
relaxation properties of 124
in 50% (w/w) solution in toluene,
C-13 NOEFs of 126*t*
in 50% (w/w) solution in toluene,
C-13 spin-lattice relaxation
times of 126*t*
Poly(*n*-butyl methacrylate)
(PBMA) 121–123
C-13 spin-lattice relaxation times of 122*t*
as a function of temperature, NT_1
and NOEF values for 135*t*
in 50% (w/w) solution in toluene-
d_8, NOEFs for 122*t*
Poly(chlorotrifluoroethylene)
(PCTFE) 207
C-13 MAS spectra of 207
fluorine-decoupled 210*f*
Poly(dA–dT)
260-nm absorbance melting curve
of 276*f*

INDEX

Poly(dA–dT) (*continued*)
  and its daunomycin complex in 1*M* NaCl solution, chemical shift of the thymidine H-3 proton in 259
  and the daunomycin · poly(dA–dT) complexes, temperature dependence
    of the adenosine H-2 resonance for 262*f*
    of the base and sugar H-1' resonances for 265*f*
    of the linewidths of the adenosine H-8 resonances for 265*f*
  duplex
    dissociation rate constants in 0.1*M* phosphate solution 248*t*
    formation, upfield shifts associated with poly(dA–dT) and 229*t*
    -to-strand transition of 222
  hydrogen bonding in duplex state for 220–221
  360-MHz correlation proton NMR spectra of 236*f*
  360-MHz proton NMR spectra of 223*f*, 225*f*
    of the thymidine H-3 proton in .. 221*f*
  melting transition of 237
    in 0.1*M* phosphate solution, chemical-shift parameters associated with 228*t*
  and its netropsin complex, chemical shift of the thymidine H-3 proton in 278*t*
  and the netropsin · poly(dA–dT) complex, temperature dependence of the sugar H-1' chemical shifts of 280*f*
  and poly(dA–dU) duplex formation, upfield shifts associated with 229*t*
  and the proflavine · poly(dA–dT) complex, proton noise decoupled 145.7-MHz NMR spectra of 258*f*
  and the proflavine · poly(dA–dT) complex(es), temperature dependence of
    the nucleic acid chemical shifts for 249*f*
    the nucleic acid and proflavine chemical shifts for 246*f*
    the thymidine H-3 resonance in 243*f*
  proton noise decoupled 145.7-MHz ^{31}P NMR spectra of 234*f*, 243*f*
  temperature dependence of
    the adenosine H-2 pyrimidine H-6 and adenosine H-8 protons in 227*f*

Poly(dA–dT) (*continued*)
  temperature dependence of (*continued*)
    the base and sugar proton chemical shifts of 224*f*
    the base and sugar proton resonances of 239*f*
    the chemical shifts and linewidths of the base resonances of .. 284*f*
    linewidth(s) of 230*f*
    the thymidine CH$_3$-5 chemical shift and the adenosine H-8 .... 245*f*
    the thymidine H-3 proton chemical shift of 238*f*
  resonance assignments of 222–226
  variation of the sugar H-1' and base proton chemical shifts of 282*f*
Polyenyl radical, precursor for 44
Polyethylene(s) (PE) 93
  absorption lineshapes for polycrystalline 195*f*
  advantages and disadvantages of C-13 NMR for characterizing 116
  backbone and side-chain C-13 chemical shifts from TMS as a function of branch length 100*t*
  C-13 spectra of benzene and 194
  C-13 spectra for polycrystalline 195*f*
  end group distribution 94, 115*t*
  high-density 93
    short-chain and long-chain branching in 96
  long-chain branching
    characteristics of 104*t*
    flow activation energies of 104*t*
    using high-field carbon-13 93–118
    and short-chain branching 115*t*
  low-density 93
    produced from a high-pressure process, C-13 NMR spectrum at 25.2 MHz of 102*f*
  molecular weight characteristics of 94
  Nuclear Overhauser effect in 112
  number average molecular weight for 115*t*
  number average molecular weight data from SEC for 115
  oxide (PEO) 202–204
    at -140°C, proton-decoupled/CP C-13 NMR spectra of 203*f*
    and rotating-frame relaxation time 202
  and PP, comparison of $T_{1\rho}$ data for 166*f*
  structural characteristics 94
Poly(*n*-hexyl methacrylate) (PHMA) .. 124
  relaxation properties of 124

Poly(*n*-hexyl methacrylate) (PHMA) (*continued*)
  in 50% (w/w) solution in toluene-d_8, C-13 spin–lattice relaxation times of 125*t*
Polymer(s)
  absorbed 1–18
  and free 2
  on oxide surfaces 1–18
  backbone motion 128–133
  C-13 spectrum at 50-MHz of polymers A, B, C, D, and E ... 105*f*–109*f*
  chains in solution, molecular dynamics of 119–145
  conformation and mobility of 1–18
  diffusion of PVAc from SiO_2 surface into bulk 16*f*
  influence of surface on mobility of .. 2
  molecule average number of long-chain branches 111
  solid
    background magic-angle spinning for C-13 NMR in 193–196
    multiple-pulse NMR of 169–191
    relaxation parameters for 211
    VT-MAS C-13 NMR studies of 193–217
  –surface interaction 2
  and temperature studies 11–14
Polymeric material, characterization of solid 147–168
Polymeric systems, pulsed NMR for studying composite 147–168
Polypropylene (PP) 209–214
  C-13 relaxation times for the methyl carbons of 213*f*
  comparison of $T_{1\rho}$ data for PE and 166*f*
  proton-decoupled CP/MAS C-13 .. 211
    spectra of 211
    at 24°C and -170°C, isotactic 212*t*
  relaxation results for 211
Polystyrene (PS) 2
  adsorption of PVAc and 17
  in the bulk state, temperature dependence of 13*f*
Polytetrafluoroethylene (PTFE) 169, 206
  backbone in the amorphous state .. 186
  chemical shift spectra of 170–180
    obtained at 259° 170
  samples of varying crystallinity, F-19 REV-8 .... 171*f*
  crystalline and amorphous lineshapes of 170–180
  and crystalline first-order transitions 180
  decomposed F-19 REV-8 chemical shift spectra of amorphous and crystalline fractions of 173*f*
  decomposed lineshapes of 174

Polytetrafluoroethylene (PTFE) (*continued*)
  lineshape changes observed for the amorphous fraction of 182
  local-order parameter for 184
  MAS C-13 spectra of
    at -72°C 208*f*
    fluorine-decoupled 208*f*
    as a function of temperature 207
  melt-recrystallized 170–180
  chemical shift lineshape changes for the crystalline fraction of 180
  decomposed chemical shift spectra of 170
  F-19 REV-8 chemical shift lineshapes 175*f*
    of the amorphous fraction of .. 178*f*
    of the crystalline fraction of .. 176*f*
  variable-temperature chemical shift spectra 169–180
  molecular dynamics of 169–191
  relaxation map of 180, 181*f*
  rotational diffusion rates of 180
  temperature dependence of 177*f*
  sample(s)
    crystallinities of several melt-processed and annealed-virgin 172*f*
  F-19 REV-8 chemical shift lineshapes of a 68% crystalline 172*f*
  temperature dependence of relaxation times of 179*f*
  and TFE/HFP copolymer, temperature dependence of the local-order parameter in the amorphous regions of 185*f*
  and a TFE/HFP copolymer, temperature dependence of the rms value of Gaussian distribution in the amorphous regions of 187*f*
Poly(vinyl acetate) (PVAc) 2, 11
  ESR spectrum(a) of 3*f*
    at the solid–liquid interface 7*f*
    adsorbed 7*f*
  and PS, adsorption of 17
  from SiO_2 surface into bulk polymer, diffusion of 16*f*
  temperature dependence of
    bulk 5*f*
    in the bulk state 12*f*
    on SiO_2 when adsorbed from CCl_4 .... 15*f*
    on SiO_2 when adsorbed at saturation coverage 15*f*
    on TiO_2 in $CHCl_3$, effect of molecular weight of 9*f*

INDEX

Poly(vinyl chloride) (PVC) 35
  bond cleavage in 40
  ESR spectrum(a) of
    and copolymer glasses 37–38
    radical, computer-simulated 45f
    radicals, dipolar interaction and
      line broadening in 44
  ESR study of intermediates in the
    photodegradation of 35–49
  mechanisms of photodegradation
    of 35–36
  plasticized 154
    at 75°C, free induction signals of 158f
    free induction signals of pure and 157f
    $T_{1\rho}$ and T_2 data for 161f
  prepared at 50°C 40
  prepared at -78°C 40
  radical at LNT, ESR stick spectrum 43f
  –TP glass at LNT, ESR spectrum
    of UV irradiated 39f
  and VCl–VCl$_2$ copolymers, ^{13}C
    {^{1}H} NMR spectra of 82f
Poly(vinylidene fluoride) (PVF$_2$) 209
  C-13 NMR spectrum of semi-
    crystalline 210f
  cross-relaxation rate for 165f
  energy exchange in 164
  as a function of temperature for
    protons, T_1, T_2, and $T_{1\rho}$ data for 165f
  /PMMA blend, cross-relaxation
    rate for 40/60 .... 165f
  T_1, T_2, and $T_{1\rho}$ data for 164
Premelting transition 231–232
Proflavine
  on addition of dC–dC–dG–dG and
    dG–dG–dC–dC, change in
    absorbance 253f
  –DNA complexes 241–256
  overlap geometry for 251–255
  · poly(dA–dT) complex(es)
    complexation shifts on formation
      of 250t
    melting transition of 242
    360-MHz proton NMR spectra
      of 244f
    phosphodiester linkage: 255–256
  poly(dA–dT) and
    proton noise decoupled 145.7-
      MHz ^{31}P NMR spectra of 258f
  temperature dependence of the
    nucleic acid chemical
    shifts for 249f
  temperature dependence of the
    nucleic acid and profla-
    vine chemical shift for 246f
  temperature dependence of the
    thymidine H-3 resonance
    in 243f

PP (see Polypropylene)
PSAN/PMMA blends as a function
  of temperature T_1, T_2, and $T_{1\rho}$
  data for 151f
Pulse sequence, Goldman–Shen 159, 163f
Pulsed NMR for studying composite
  polymeric systems 147–168

R

Radical I 35, 38–44
Radical II 44
Relaxation
  data for normal alkanes vs. tem-
    perature, rotating-frame 150f
  map of PTFE 180, 181f
  measurements, NMR 147
  parameters for solid polymers 211
  properties
    of DD 124
    of DT 124
    of HDD 124
    of PBA 124
    of PHMA 124
    of side-chain carbons 133
  results for PP 211
  spin–lattice 150f
  spin–spin 150f
  time(s)
    for the methyl carbons PP, C-13 213f
    PEO and rotating-frame 202
    of a PTFE sample, temperature
      dependence of 179f
    spin-lattice
      for the C-1 carbon, methylene 141t
      for the C-2 carbon, methylene 142t
      C-13 119
      of PBMA 122t
      of PHMA in 50% (w/w)
        solution in toluene-d_8 .. 125t
      for neat
        DD 127t
        DT, C-13 .... 131t
        HDD, C-13 .... 130t
    spin–spin 119
Resonance(s)
  assignments of poly(dA–dT) 222–226
  carbon 101
  mutagen 247
  nucleic acid base 242–247
  nucleic acid sugar 247
Rotating-frame relaxation data for
  normal alkanes vs. temperature 150f
Rotating-frame relaxation time, PEO
  and 202
Rotational correlation time of the
  enzyme-Gd^{3+} complex 71
Rotational diffusion rates of PTFE 180

S

Sarcoplasmic reticulum Ca^{2+}-ATPase, Gd(III) as a probe in NMR and EPR studies of 64–74
Scanning electron micrographs (SEM) 23
Segmental mobility
  in the dry state, effect of surface on 11–17
  dynamics 14–17
  of loops in the solid state 14
Sequence specificity 247–251
  for the daunomycin · poly(dA–dT) complex 268
SEM (see Scanning electron micrographs) 23
Size exclusion chromatography (SEC) 116
  of Hizex 7000 ..... 95f
  of NBS 1475 ..... 95f
  for PEs, number average molecular weight data from 115
  of Phillips PE 5003 ..... 95f
SiO_2(Cab-O-Sil)-CCl_4 interface, dependence of the ESR spectra on surface coverage of PVAc at 10f
Solid–liquid interface
  conformation and mobility at 6–14
  effect
    of molecular weight 8
    of polymer on 8
    of solvent on 6–8
    of surface 6–14
      coverage on 8
  ESR spectra of PVAc at 7f
Spin
  diffusion 148
    to a sink, model calculation for 156f
  –lattice relaxation time(s) 150f
    for distribution widths 134f
    for the C-1 carbon, methylene 141t
    for the C-2 carbon, methylene 142t
    C-13 119
    for neat DT 131t
    for neat HDD 130t
    of PBA in 50% (w/w) solution toluene 126t
    PBMA 122t
    of PHMA in 50% (w/w) solution in toluene-d_8 ..... 125t
    for the log-χ^2 distribution 132f
    for neat DD 127t
  –spin relaxation 150f
    time 119
    temperature 148
Spinner assembly–probe geometry used for VT-MAS 203f
Spinning apparatus, spinning rates vs. applied gas pressures for 200f
Sugar proton complexation shifts for the daunomycin · poly(dA–dT) complex 264

Surface degradatoin, effect between environment and stress 32
Surface on segmental mobility in the dry state, effect of 11–17

T

$T_{1\rho}$ decay of 75/25 PSAN/PMMA 155f
Tetrahydrofuran (THF) 37
Tetrahydropyran (TP) 37
TFP/HFP copolymer, temperature dependence of the local-order parameter in the amorphous regions of PTFE and 185f
TFE/HFP copolymer, temperature dependence of the rms value of Gaussian distribution in the amorphous regions of PTFE and 187f
Thymidine
  proton chemical shift of poly(dA–dT), temperature dependence of 238f
  proton in poly(dA–dT) and its daunomycin complex in 1M NaCl solution, chemical shift of 259t
  360-MHz proton NMR spectrum of 221f
  and its netropsin complex, chemical shift of 278t
  resonance in poly(dA–dT) and the proflavine · poly(dA–dT) complex, temperature dependence of 243f
  resonances, temperature dependence of the linewidth of the nonexchangeable adenosine H-8 and 238f
Transition
  midpoints of the daunomycin · poly(dA–dT) complexes 263t
  of the poly(dA–dT), melting 237
  premelting 231
Transverse relaxation rate 51

V

Variable-temperature magic-angle spinning (VT-MAS) 193
  CPBD and 204
  fluoropolymers and 206–209
  spinner assembly–probe geometry used for 203f
Vinyl chloride–vinylidene chloride (VCl–VCl_2) copolymers
  carbon environment in 83f
  -13 chemical shifts for 85

Vinyl chloride–vinylidene chloride
 (VCl–VCl$_2$) copolymers (*continued*)
 -13 NMR
  analysis of 87t
  areas, calculation of monomer
   composition of 86
  molecular structure of 81–91
  ^{13}G {^{1}H} NMR spectra of PVC and 82f
 Vinylidene chloride (VCl$_2$) monomer 81
  isolated 84
  number average sequence lengths
   for 88–90

Vinylidene chloride (VCl$_2$) monomer
 (*continued*)
  paired 84
  greater than 84
VT-MAS (*see* Variable-temperature
 magic-angle spinning)

Z

Zeeman field 209
Zero field splitting (ZFS) 50